W0257714

Sergej A. Akhmanov was born in 1929 in Moscow,
U.S.S.R. He graduated in 1952, and received the
Ph.D. degree in physics in 1964, and the Doctor of
Sciences degree in 1968 all from Moscow State Uni--
versity, Moscow, U.S.S.R.
He has been a Professor with the Department of Phy-
sics and the Chairman of the Chair of General Phy-
sics and Wave Processes, Moscow State University,
Moscow, since 1974.
Since 1977 he has been the head of the Laboratory
of Nonlinear Optics of MSU.
Prof. S. A. Akhmanov is the coauthor of a number of
scientific monographs.

Vladimir I. Emel'janov was born in 1943 in Moscow.
He graduated in 1966 and received Ph.D. degree
in Physics in 1973 and the Doctor of Science degree
in 1989 all from Moscow State University.
He is now a senior research stuff member with the d
partment of Physics, Moscow State University. His
field of research is coherent cooperative phenomena
in optics, nonequilibrium phase transitions and interaction of powerfull laser radiation with surface
and interfaces. He is coauthor of a monograph.

Nikolai I. Koroteev was born in 1947 in Volgograd,
U.S.S.R. He graduated in 1971, received Ph.D. degre
in physics in 1974, and the Doctor of Sciences degr
in 1983, all from Moscow State University, Moscow,
U.S.S.R.
He is now a full professor with the Department of
Physics, and the director of International Laser Ce
ter Moscow State University, Moscow, U.S.S.R.
His field of research is the application of nonline
ar optical spectorscopy techniques in the fields of
molecular physics, optics of surfaces, excited stat
etc. He is coauthor of a number of monographs.

Akhmanov, Sergej, A.: Interaction of Strong
Laser Radiation with Solids and
Nonlinear Optical Diagnostics of
Surfaces / Sergej A. Akhmanov, Vladimir I.
Emel'yanov, Nikolaj I. Koroteev. - Leipzig
BSB Teubner (Teubner-Texte zur Physik; 24) -
1. Aufl. - 1990. - 204 S.
NE: Emel'yanov, Vladimir I.; Koroteev,
Nikolaj I.; GT

ISBN 978-3-322-96426-7   ISBN 978-3-322-96425-0 (eBook)
DOI 10.1007/978-3-322-96425-0

ISSN 0233-0911
TEUBNER-TEXTE zur Physik · Band 24
© BSB B. G. Teubner Verlagsgesellschaft, Leipzig, 1990
1. Auflage
VLN 394-375/79/90 · LSV 1165
Lektor: Dr. rer. nat. Ingrid Lerchner

Gesamtherstellung: (52) Nationales Druckhaus, Berlin

TEUBNER-TEXTE zur Physik · Band 24

Herausgeber/Editors: Werner Ebeling, Berlin

Wolfgang Meiling, Dresden

Armin Uhlmann, Leipzig

Bernd Wilhelmi, Jena

S. A. Akhmanov / V. I. Emel'yanov / N. I. Koroteev

Interaction of Strong Laser Radiation with Solids and Nonlinear Optical Diagnostics of Surfaces

The results on the experimental and theoretical investigations of the processes of strong laser light-surface interaction and on the nonlinear optical diagnostics of these processes and of the surface states are given. The second harmonic generation and picosecond CARS in reflection and other optical methods permit to obtain valuable structural information with high temporal and spatial resolution. Two new classes of laser-induced surface instabilities - the interferential instability of the surface relief and the diffusional-deformational instability are considered in detail.

Gegenstand des Buches sind die Ergebnisse aus der experimentellen und theoretischen Erforschung der Prozesse der Wechselwirkung zwischen einer starken Laserstrahlung mit der Oberfläche stark absorbierender Medien sowie die aus der optisch nicht linearen Diagnostik dieser Prozesse und der verschiedenen Oberflächenzustände. Das Generieren der zweiten Harmonischen auf die Reflexion, das Pikosekunden – CARS auf die Reflexion und die anderen optischen Methoden liefern aufschlussreiche Strukturinformationen mit einem hohen räumlichen und zeitlichen Auflösungsvermögen. Eingehend behandelt sind zwei neue umfangreiche Klassen laserinduzierter Oberflächeninstabilitäten – die Interferenzinstabilitäten des Oberflächenreliefs und die Diffusions- und Deformationsinstabilitäten.

Dans le livre sont exposés les résultats de l'étude théorique et expérimentale des processus de l'action du puissant rayonnement du laser a la surface des milieux fortement absorbants, aussi bien les résultats du diagnostique optique nonlinéaire de ces processus et les différents états de la surface. La génération de la deuxième harmonique sur la réflexion, le CARS picoseconde sur la réflexion et d'antres méthodes optiques donnent une riche information structurale à haut degrée de resolution temporelle et spatiale. Sont examinées de nouvelles classes des instabilités sur la surface – les instabilités interférentielles du relief de la surface et les instabilités déformationnelles.

El libro da a conocer los resultados del estudio experimental y teórico de los procesos verificados bajo la acción recíproca entre una potente radiación de láser y la superficie de medios altamente absorbentes, así como de referidos procesos y de diferentes estados de la superficie. La generación del armónico secundario sobre la reflexión, el CARS pico-segundo sobre la reflexión y otros métodos ópticos proporcionan una vasta información estructural de alto grado de solución espacial y temporal. Fueron detalladamente estudiadas dos nuevas clases de inestabilidades inducidas por láser sobre la superficie: inestabilidades interferenciales del releive de la superficie e inestabilidades de deformación difusivas.

В книге изложены результаты по экспериментальному и теоретическому изучению процессов взаимодействия мощного лазерного излучения с поверхностью сильно поглощающих сред и нелинейно-оптической диагностике этих процессов и различных состояний поверхности. Генерация второй гармоники на отражение, пикосекундный КАРС на отражение и другие оптические методы дают ценную структурную информацию с высоким пространственным и временным разрешением. Подробно рассмотрены два новых широких класса лазерно-индуцированных неустойчивостей на поверхности – интерфереционные неустойчивости рельефа поверхности и диффузионно-деформационные неустойчивости.

2

CONTENTS

Foreword 7

Introduction 8

1. **Absorbtion and relaxation of laser pulse energy. Lattice heating** 12
1.1. Impulsive laser excitation and relaxation of the electronic sybsystem 13
1.2. Electron-phonon relaxation and lattice heating 18
1.3. Pulsed laser annealing of a semiconductor surface. Thermal and
plasma annealing mechanisms 21

2. **Linear and nonlinear optical methods for recording the states and
laser-induced phase transitions on the surface of condensed media** 26
2.1. New data from linear optical spectroscopy on fast laser-induced
processes on semiconductor and metal surfaces 26
2.1.1. The study of the optical reflection from surfaces in real time in
pulsed laser annealing regime 26
2.1.2. Raman and stimulated Brillouin scattering of light by the laser-
excited surfaces of semiconductors 29
2.1.3. Optical recording of periodic surface ripples 31
2.2. Nonlinear optical diagnostics of laser-induced transitions on
a semiconductor surface; generation of optical harmonics and
combination frequencies in reflection 31
2.2.1. Nonlinear optical diagnostics of the surface 31
2.2.2. Theory of generation of optical harmonics in reflection. Nonlinear optical source 33
2.2.3. Field of reflected second and third harmonics generated by nonlinear source 38
2.2.4. Surface and bulk nonlinearities contributing into the reflected
harmonic generation 42
2.3. Study of the states and fast phase transformations of semiconductor surfaces by means of the generation of optical harmonics
and sum frequencies on reflection 45
2.3.1. Experiments on second harmonic generation in reflection for
the study of the pulsed laser annealing dynamics in GaAs 52
2.3.2. Diagnostics of quality of laser annealing on basis of second
harmonic generation in reflection 52

2.3.3.	Determination of degree of disordering of a crystal surface from third harmonic and second harmonic generation in reflection ...	54
2.3.4.	Subpicosecond dynamics of structural changes of Si surface in pulsed laser annealing ...	63
2.3.5.	Nonlinear optical probing of stressed and deformed interfacial layers ...	64
2.3.6.	The dynamics of Si lattice on a picosecond time-scale; picosecond CARS in reflection ...	67
3.	**Nonlinear optical process determined by the modulation of the surface relief and of the surface temperature of condensed media** ...	72
3.1.	Laser-induced instabilities of the surface relief and the problem of the surface periodic structures (ripples) formation ...	72
3.1.1.	Optical exitation of surface periodic structures ...	72
3.1.2.	The physics of rough surface-laser field interactions ...	75
3.1.3.	The feedback mechanisms and evolution of structures ...	77
3.2.	Diffraction of light waves by small spatial-temporal modulations of the surface relief ...	79
3.3.	The feedback mechanisms in premelting regimes. The generation of coupled surface electromagnetic and acoustical waves ...	83
3.3.1.	The exitation of the surface temperature wave owing to the interference of surface electromagnetic wave and transmitted laser wave ...	83
3.3.2.	The laser generation of coherent surface acoustical waves and statical gratings ...	85
3.3.3.	The surface acoustical waves characteristics in dependence on the pump wave parameters ...	90
3.4.	The capillary waves and evaporation waves generation under the action of laser radiation on the liquid metals, semi-conductors and dielectrics ...	94
3.4.1.	Spatially nonuniform heating, evaporation and motion of liquid under the action of laser radiation ...	95
3.4.2.	Dispersion equation for relief modulation waves. Competition between three feedback machanisms in the process of surface structures formation in different laser heating regimes ...	100
3.4.3.	Generation of capillar waves and the characteristics of the dominant surface structures for s-polarization of the pumping wave .	103
3.4.4.	Generation of capillar waves and the characteristics of the dominant surface structures for p-polarization of the pumping wave	108

3.4.5.   Nature of the doublet structures 110

3.4.6.   Generation of small-scale structures under the action of laser radiation 113

3.5.   Dependence of the orientations and periods of the surface ripples generated due to the interference instability on the dielectric permeability of the medium 115

3.5.1.   s-polarization of the pump wave 115

3.5.2.   p-polarization of the pump wave 123

3.6.   The nonlinear regime of the laser generation of capillar waves and the formation of surface ordered structures 126

3.6.1.   The diffraction of the light on the periodically modulated surface relief with large amplitude of modulation 127

3.6.2.   The equations for the nonlinear regime of capillar waves generation . . . 129

3.6.3.   The dispersion equation for determining the stationary amplitudes of laser induced capillar waves 132

3.6.4.   The stationary frequencies and amplitudes of the laser induced capillar waves 134

3.6.5.   The multimode regime of capillar wave generation and the formation of the hexagonal surface structures 140

3.7.   The self-induced strong enhancement of the meterial absorptivity under the surface ripples generation 144

3.7.1.   The dynamics of the laser heating of the surface with the periodically modulated relief 145

3.7.2.   The enhanced absorptivity and rapid heating of the surface for s-polarization of the pump 147

3.7.3.   The enhanced absorptivity and rapid heating of the surface by p-polarized pump wave 148

3.7.4.   The discussion and the comparison with the experiment 150

3.8.   The comparison of the linear theory of the interferential instabilities of the surface relief and the experimental results on ripples generation 151

4.   The diffusional-deformational instabilities and phase transitions on the surface under the action of laser radiation 157

4.1.   The electron-deformational-thermal instability and the semiconductor-metal phase transition with the formation of the ordered surface structures 157

4.1.1.   The closed system of equations for the modulations of the density of the nonequilibrium carriers, the temperature and the vector of the displacement of the medium 161

4.1.2.   The general dispersion equation of the electron-deformational-thermal instability. The one dimensional gratings formation   163

4.1.3.   The nonlinear regime of the electron-deformational-thermal instability and semiconductor-metal phase transition   170

4.1.4.   The formation of radial ringed structures   171

4.1.5.   The comparison with the experiment and conclusions   173

4.2.   Laser-induced defect-deformational instabilities and formation of the ordered defect structures   175

4.2.1.   Dislocational-deformational instability and formation of the dislocational gratings under the action of laser light   175

4.2.2.   The laser induced void-deformational instability and formation of the ring structures of the voids in the thin films   177

4.3.   Crystallizational-deformational-thermal instability and the formation of the ordered crystalline-amorphous structures under laser crystallization   178

Conclusions   187

References   190

Appendix   201

6

FOREWORD

Present monograph is devoted to the nonlinear optical processes and nonlinear optical diagnostics of pulsed laser-matter interactions in subsuface layers. This is very hot field now and many scientists from all over the world are involved in this research area. We hope this monograph will contribute into the common efforts to cover this field and to create new opportunities both in fundamental research and practical applications.

The list of the most probable areas of practical technological applications of processes, covered by this book, includes laser material processing, laser probing of microelectronic devices, laser control of technological processes at surfaces, etc. We hope that book will be of interest not only for the laser specialists, but also for engineers and students, working in the field of practical applications of laser-matter interactions.

The last years of research activity have brought many new fruitful ideas and valuable results in the field of pulsed laser radiation interaction with the surfaces of solids and liquids. The main experimental achivements, which have created the conditions for the rapid progress in this research area are the development of a new generation of pulsed lasers, emitting femtosecond optical pulses and the development of new probing techniques based on the nonlinear interaction of probing pulses with surface under investigation. This technique is possessed of a good temporal and spatial resolution as well as a high enough sensitivity. The theoretical efforts in the last years have lead to more compehensive appreciation of the fundamental role of surface instabilities during the intense pulsed laser action on metal, semiconductor and dielectric materials. Now we understand the nature and mechanism of a number of different instabilities much better and deeper than three or four years ago.

The main source of detailed information for this book are the results of recent research obtained in R.V.Khokhlov's laboratory of nonlinear optics of Moscow State University. We acknowledge with many thanks the fruitful collaboration with Drs. I.L.Shumay, S.V.Govorkov, G.A.Paitian, V.N.Zadkov, V.N.Seminogov, V.V.Yakovlev, G.I.Petrov, A.Yu.Abdullaev and extremely usuful discussions with Drs. V.N.Bagratashvili, V.Ya. Panchenko, P.K.Kashkarov and A.A.Sumbatov.

S.A.Akhmanov, V.I.Emel'janov, N.I.Koroteev
Moscow, December, 1988.

INTRODUCTION

The aim of the present book is to survey the new developments in the field of strong laser beam-solid interactions, which encompasse a large group of physical phenomena associated with the excitation of strongly non-equilibrium states in the surface layers of metals, semiconductors and dielectrics. The special emphasis is made on the role of nonlinear response of materials in these phenomena and diagnostics.

The absorption of laser radiation which is strongly concentrated in space and time gives rise to the entire cascade of energy transformation processes, which involve the successive excitation and relaxation of electron subsystem, electron-phonon relaxation, phonon-phonon relaxation and finally various thermal processes (heating, melting and evaporation).

The final thermal stage in metals has been studied for at least 20 years (see, for example, [1–4]); experiments with laser pulses with the duration from 10^{-3} to 10^{-6}s give practically exhaustive physical information about this stage. The thermal effect of the absorption of laser radiation on metals has opened up a wide field of applications; among them are laser cutting, welding, hardening, drilling and laser thermochemical treatment of metals.

The results of similar experiments with strongly absorbing semiconductor crystals have proved to be extremely interesting. The action of high power laser pulses results in a rapid and high quality recrystallization of the surface layer of the semiconductors which have been rendered amorphous by ion implantation i.e. pulsed laser annealing (PLA) occurs [5–12]. In spite of the fact that the laser annealing has already become a well developed technological tool, experiments on laser annealing put forward a number of still unsolved important physical problems. The work carried out in the last 3–5 years have shown that fast laser-induced phase transformations in surface semiconductor layers (melting-solidification, amorphous solid-crystal and crystal-amorphous solid phase transitions) occur on nano-, pico- and even subpicosecond time scales. For detailed explanation of these diverse and in many ways unexpected phenomena it is necessary to give answers to a number of fundamental questions about the behaviour of semiconductors in a strong laser field.

What is the nature and how high are the rates of electronic electron-phonon and phonon-phonon relaxation processes when intense laser field generates up to 10^{22} free carriers per cm^3, highly excited above the bottom of conduction band within times of $10^{-9} - 10^{-14}$s? What is the state of such a dense laser-induced electron-hole plasma? How and through what stages does melting of a crystal, containing this dense hot plasma, proceed: through direct transfer of free carriers excess energy to lattice vibrations or through plasma-induced soft phonon modes? Does the crystal melting occurs after the energy, obtained from the electronic subsystem, is thermalized among all the phonon modes or when it is concentrated only in short wavelength part of phonon spectrum while phonon modes away from the center of the Brilluoin zone remain unexcited?

Existing theories provide different, and sometimes contradictory answers to these questions, so that adequate experimental diagnostics plays a fundamental role in obtaining unambiguous conclusions.

At the present time most experimental data support simple "thermal" mechanism of pulser laser annealing of elementary group IV and $A^{III} B^{V}$ compound semiconductors for laser pulses with the duration of several picosecond and longer (see [11, 12] and Sec.1). According to this model, laser pulse annealing is accomplished through a very fast (characteristic times ~ 1 ps) heating of the crystal lattice owing to electron-phonon relaxation and subsequent fast melting of crystal surface layer. After the action of laser pulse and cooling of the surface the first-order phase transition either to crystalline, or to poly-crystalline, or to solid amorphous state takes place. There are, however, experimental data which do not fit into the "thermal" model. Therefore it is very important that at present it has become possible to investigate experimentally the kinetics of all stages of the process of laser pulse energy transformation in semiconductors and metals on time intervals as small as 10^{-14}s and to give clear-cut answers to the questions posed above.

This is linked to the creation of stable frequency-tunable femtosecond pulse lasers (1 fs $-$ 10^{-15}s) [13—15], every pulse envelope encompassing only a few cycles of the optical field, perfectly fitted for the purposes of optical diagnostics techniques. The significance of these achievements, of course, extends far beyond the technological requirements of laser annealing.

Probably the most characteristic feature of the present-day study of laser-induced nonequilibrium states and laser-induced phase transition is the clear understanding of the important and often decisive role, played by the **nonlinearity of matter response** in the phenomena under consideration.

In laser thermochemistry, for example, a wide class of nonlinear processes resulting in the formation of spatially and temporally ordered structures has been recently studied in detail [17].

Strongly dependent on the excitation level (and, hence, nonlinear) is the relaxation of non-equilibrium states created by high-power laser pulses. Such non-linear relaxation, quite through investigated in polyatomic molecules [21] is ever more often encountered in condensed media [22—24]. Finally, nonlinear spectroscopy methods provide information on the development of excitation and relaxation processes on a real time scale (including down to times 10^{-14}s).

The generation of optical harmonics and combination frequencies "in reflection" [17—19] are very effective here; the active spectroscopy of light scattering by plasma oscillations, optical and acoustical lattice vibrations is also promising [21,214]. Thus, **nonlinear excitation, nonlinear relaxation, nonlinear-optical diagnostics** of nonequilibrium processes have actually become key problems of the physics of high-power laser beam-solid interactions.

Spontaneously arising different instabilities at surface, giving rise to the formation of surface periodical structures (SPS) are another vivid example of the role of optical nonlinearity in the process of laser excitation of semiconductor, dielectric and metal surfaces.

The formation of SPS on the surface of solids irradiated by laser light is an universal phenomenon, spontaneously arising whenever laser light intensity is sufficiently high. The qualitatively different SPS are formed owing to the development of the surface instabilities of two different classes.

To the first class belong the interferential instabilities of the surface relief with surface periodical ripples formation. The origin of this phenomenon is linked to the spatially nonuniform surface heating; the required nonuniform field is formed by the interference of the incident laser wave with the field formed by the laser radiation scattering (diffraction) by initially random modulations of surface relief. Of special physical interest is the possibility of positive "feedback", when the periodic structures formed begin to affect substatially the laser radiation scattering in the diffracted waves. The resultant instability is similar to the one caused by stimulated scattering. Under different conditions unstabilities of surface acoustic waves (SAW), capillary waves (CW) in melts and liquid metals and, finally, interference evaporation instability (IEI) of surface can take place.

In the laser-induced instabilities of the second class — the diffusional-deformational instabilities (DDI) the deformation of the medium modulates the activation energy of some diffusional (or kinetical) variable (concentration of photo-exited carriers, vacancies, dislocations, voids). In its turn, the diffusional variable in the presence of laser field turns out to be coupled to the deformation, which leads to the instability at some threshold value of laser intensity. As a result of DDI the SPS of diffusional variable coupled to the deformational field are formed.

The main emphasis of the present book is made on the problems mentioned above in the cases of semiconductors, dielectrics and metals. The book consists of four sections. In the first one we shall briefly review the basic ideas concerning the fundamentals of laser pulse energy absorption relaxation and the heating of lattice. The second section is devoted to discussion of the linear and newly developed nonlinear method of diagnostics of surface states and ultrafast laser induced surface processes. In the third section the detailed exposition of the linear and nonlinear theory of the interferential instabilities and ripples formation in comparison with the experimental results is given. At last, the fourth section deals with the diffusional-deformational instabilities and diffusional-deformational ordered structures formation.

1. ABSORPTION AND RELAXATION OF LASER PULSE ENERGY. LATTICE HEATING.

If a semiconductor is subjected to laser radiation with the photon energy $\hbar\omega$ substantially exceeding the gap width E_g: $\hbar\omega > E_g$ (Fig. 1), light absorption occurs in a thin surface layer with thickness $\gamma^{-1} \sim 10^{-4} - 10^{-6}$ cm, where γ is the optical absorption coefficient. (The latter value is also characteristic for the absorption depth of metals). Under pulsed laser action of this kind, when radiation intensity is high ($I \sim 10^6 - 10^{12}$ W/cm^2), a strongly nonequilibrium and non-stationary state of both electronic and phonon subsystems is created in the layer.

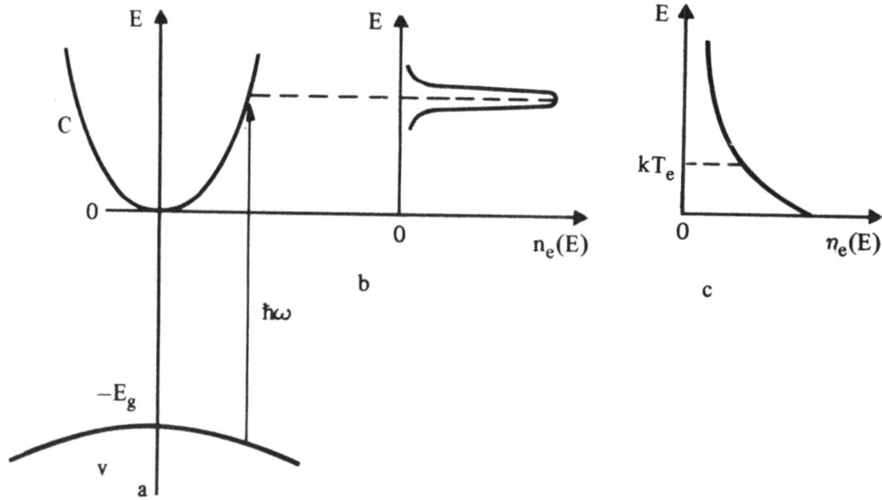

Fig. 1. Laser excitation and intraband energy relaxation of carriers in a semiconductor: a) interband transition with absorption of a quantum $\hbar\omega$; b) initial form of the electron distribution function; c) final form of the energy distribution function.

Since the primary energy absorption occurs in the electronic system, initially there is a substantial difference between electron temperature T_e and lattice temperature T. The process of energy transfer and thermalization in the lattice includes a series of relaxation stages both within the electron-hole subsystem and electron-phonon and phonon-phonon relaxation.

In the present section we shall consider in more detail each of the mentioned stages of laser-radiation interaction with semiconductor surface and subsequent processes.

1.1. Impulsive laser excitation and relaxation of the electronic sybsystem

If a semiconductor surface is subjected to laser pulsed radiation with energy density W, duration τ_p and photon energy $\hbar\omega > E_g$, owing to interband absorption (see Fig. 1), nonequilibrium electron-hole pairs are created at a rate, which can be determined from the following equation:

$$G(t,z,T) = \frac{\gamma(t,z,T)(1-R)}{\hbar\omega\,\tau_p}\, W \exp\left(-\int_0^z \gamma(t,z',T)dz'\right) \approx G_0 \exp(-\gamma z); \qquad (1.1)$$

here $\gamma = \gamma_l + \gamma_{nl}$ is the sum of linear and nonlinear absorption coefficients, depending on coordinate z (the axis z is directed into the medium perpendicular to the surface), T is the temperature; $R = R(t,T)$ is the optical reflection coefficient. The values of optical parameters entering into (1.1) for crystalline and amorphous silicon (Si, E_g = 1.12 eV) and gallium arsenide (GaAs, E_g = 1.43 eV) at certain frequencies, corresponding to wavelengths of pulsed lasers used in laser annealing are listed in Table 1. It is evident that the rate of laser-induced generation of free carriers can reach great values $G_0 = 10^{30} - 10^{35}$ cm^{-3}/s (for W = 0,1 J/cm^2 and $\tau_p = 10^{-8} - 10^{-13}$ s).

Photoexcited electrons have an energy of $\hbar\omega - E_g$ (zero of the energy is at the bottom of the conduction band) and initially strongly nonequilibrium distribution function centered near the energy value $E = \hbar\omega - E_g$ (see Fig. 1). Holes ($\hbar$) have analogous distribution.

The nature of the subsequent relaxation of nonequilibrium carriers in the energy and coordinate spaces depends substantially on the carrier density in the photoexcited electron-hole plasma $n_c (n_c = 2n_e = 2n_h)$. To estimate the unitial concentration of nonequilibrium carriers created by the picosecond and phemtosecond pulses we shall neglect the diffusion and carrier recombination

$$n_c \lesssim \left(\frac{dn_c}{dt}\right)_{gen} \tau_p \approx G_0 \tau_p \approx \begin{cases} 10^{22} \text{ cm}^{-3}, & \tau_p = 30 \text{ ps}, \\ \\ 10^{21} - 10^{22} \text{ cm}^{-3}, & \tau_p = 90 \text{ fs}. \end{cases}$$

Then, using Table 1, we obtain the following estimates of the upper limit:

Table 1.

Typical characteristics of laser pulses and semiconductor crystals
used in impulsive laser annealing

Laser	Wave-length λ, nm	Pulse duration τ_p, s	Coefficient of inter band absorption at T_0=300K, γ, cm^{-1}*		Coefficient of reflection 300K, R*		Rate of generation of free carriers with W=0,1 $\frac{J}{cm^2}$, G_0, cm^{-3}/S	
			Si	GaAs	Si	GaAs	Si	GaAs
Second harmonic	532,1	$1,5\ 10^{-8}$	$1,25\ 10^4$	$8,09\ 10^4$	0,37	0,38	$1,4\ 10^{29}$	$0,9\ 10^{30}$
YAG : Nd		$2\ 10^{-11}$					$1,0\ 10^{32}$	$6,4\ 10^{32}$
Third harmonic	354,7	$1,5\ 10^{-8}$	$1,07\ 10^6$	$7,14\ 10^5$	0,57	0,42	$0,5\ 10^{32}$	$0,5\ 10^{31}$
YAG : Nd		$2\ 10^{-11}$					$0,37\ 10^{34}$	$0,37\ 10^{34}$
Fourth harmonic	266,0	$1\ 10^{-8}$	$2,08\ 10^6$	$1,61\ 10^6$	0,74	0,55	$0,7\ 10^{31}$	$0,9\ 10^{31}$
YAG : Nd		$1,5\ 10^{-11}$					$0,46\ 10^{34}$	$0,6\ 10^{34}$
Dye	620,0	$0,9\ 10^{-13}$	$4,47\ 10^3$	$4,28\ 10^4$	0,35	0,35	$1\ 10^{34}$	$0,9\ 10^{35}$
Ruby	694,3	$3\ 10^{-8}$	$2,35\ 10^3$	$2,7\ 10^4$	0,34	0,34	$1,8\ 10^{28}$	$2\ 10^{29}$

*Aspnes D. E., Studna A. A. – Phys. Rev. B, 1983, v. 27, p. 985.

14

For nanosecond pulses it is neccessary to take into account the recombination. At high nonequilibrium carrier density nonradiative Auger recombination dominates, in which processes an electron and a hole, recombine, giving up their energy to a third carrier (see Fig. 2 and [22, 23]). Recombination rate then is given by $R_c = Cn_c^3$, where $C = \text{Const}$ (for example, for silicon $C = 4 \cdot 10^{-31}\,\text{cm}^6/\text{s}$). This is one examples of the nonlinear relaxation (depending on the excitation intensity) in a strongly excited electron-hole subsystem of the crystal, described in the Introduction. We shall determine recombination time by the formula

$$R_c = \frac{n_c}{\tau_r}, \qquad \tau_r = \frac{1}{Cn_c^2} \tag{1.2}$$

(It should be noted that screening of Coulomb interaction in a dense plasma limits the decrease of τ_r with increasing n_c. Thus, according to theoretical estimates, τ_r decreases assymptotically to the value $\tau_r = 6$ ps, when n_c exceeds the value $n_c = 10^{21}$ cm^{-3}).

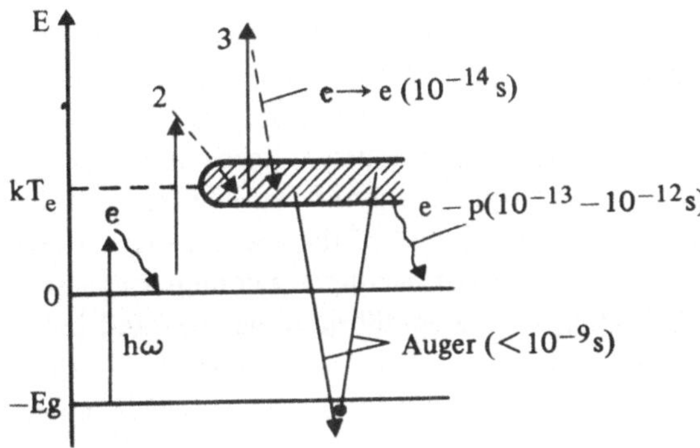

Fig. 2. Schematic energy diagram of electronic transitions in a semiconductor under the conditions of intense laser irradiation. 1) Interband absorption; 2, 3) absorption of light by free carriers. The letters e−e and e−p denote the electron-electron and electron-phonon collisions; Auger indicates Auger processes.

Stationary carrier concentration state is established for sufficiently powerfull pulses with $\tau_p > \tau_r$. From the equation for nonequilibrium carrier density

$$\frac{\partial n_c}{\partial t} = D_c \frac{\partial^2 n_c}{\partial z^2} + G - R$$

(D_c is the diffusion coefficient), putting $\partial n_c/\partial t = 0$ and neglecting diffusion, we obtain R = G, whence, using (1.2) and Table 1, we have with τ_p = 15 ns in Si and λ = 532 nm:

$$n_c < G_0 \tau_r \approx 10^{20} \text{ cm}^{-3}$$

$$\tau_r \approx 0.5 \cdot 10^{-9} \text{ s} .$$

This estimate is valid, if the distance over which the carrier diffuses before recombining $l_r = (D_c \tau_r)^{1/2} < \gamma^{-1}$, where γ^{-1} is the absorption length. In this example $D_c \sim 10$ cm^2/s and $l_r \cong 0.7 \cdot 10^{-4}$ cm $\sim \gamma^{-1}$.
The above made estimates show that for $G_0 > 10^{30}$ cm^{-3}/s the nonequilibrium carrier density under the conditions typical for laser annealing can easily exceed the value $n_c = 10^{19}$ cm^{-3}. Further we shall refer to these conditions as high excitation level regime.
At such high carrier densities their collision frequency $\tau_{e-e}^{-1} \approx \tau_{e-h}^{-1} \sim n_c$ ($\tau_{e-e}^{-1} \approx 10^{14}$ s^{-1} with $n_c \approx 10^{19}$ cm^{-3}) starts to exceed the emission rate of longitudinal (LO) and transverse optical (TO) phonons $\tau_{e-TO}^{-1} \sim \tau_{e-LO}^{-1} \sim 10^{14} -$ $- 10^{13}$ s^{-1}, which does not (or only slightly) depends on n_c [26]. Thus, intraband energy relaxation in high excitation level regime occurs over a time of the order of τ_{e-e} owing to interparticle collisions, conserving the total energy of carrier subsystem. Besides, as have been already noted, Auger recombination processes also conserve the total energy of carrier system in this regime. Due to this two circumstances practically all of the absorbed laser pulse energy at times of the order of τ_{e-e} remains within the semiconductor plasma subsystem and is "thermalized", leading to the equilibrium electron and hole energy distribution functions, characterized by the same temperature values $T_e = T_h = T_c$ (Fig. 1). The latter quantity depends on the absorbed energy and can considerably exceed lattice temperature T (which at these times remains practically equal to the initial temperature T_{in}). The value T_c, naturally, depends on the density n_c, and for $n_c \approx 10^{21}$ cm^{-3} the hot carrier temperature can reach values of $T_c \sim 10^4$K [24, 27].
Thus, in high level excitation regime at times t $\sim \tau_{e-e} \sim 10^{-14}$s the semiconductor is characterized by the presence of extremely hot ($T_e \sim 10^4$K) and dense ($n_c \sim 10^{21}$ cm^{-3}) plasma and cool lattice (T $\sim T_{in} \approx 300$ K).
In the case of metals the similar effect of anomalous electronic subsystem heating (with phonon subsystem remaining cool) was discussed theoretically in [68]. Recently the experiments have been reported, in which the electron-phonon nonequilibrium state was recorded (though indirectly) by using femtosecond laser pulses [69].

16

At low excitation levels ($n_c < 10^{19}$ cm^{-3}) optical phonon emission time in a semiconductor $\tau_{e-LO} < \tau_{e-e}$ and the excited carrier energies relaxation occurs over a time $\tau_{e-LO} \sim 10^{-13}$c^{-1} owing to optical phonon emission (see [189, 259]).

As was noted in the Introduction the use of femtosecond pulse lasers allowed the direct observation of energy relaxation from photoexcited nonequilibrium carrier distribution [13, 15]. In the paper [28] the times τ_{e-LO} and τ_{e-e} in the semiconductor $Al_x Ga_{1-x} As$ (x =0.34) were measured. The results of this experiment are given in Fig. 3 (the excitation intensity of the order 1 mW, as estimated by the authors [28] corresponds to the carrier density of the order of 10^{19}cm^{-3}). It is seen, that when $n_c < 10^{19}$cm^{-3} the main contribution to the effective relaxation constant $\tau_e^{-1} = \tau_{e-e}^{-1} + \tau_{e-LO}^{-1}$, is made by electron-phonon collisions, i.e. intraband relaxation time $\tau_e \approx \tau_{e-LO}$ and is independent of n_c, while for $n_c > 10^{19}$cm^{-3} interparticle collisions make the main contribution and $\tau_e \approx \tau_{e-e} \sim 1/n_e$. The experimental results [28] obtained for $Al_x Ga_{1-x} As$ can be applied, to a certain extent, also to GaAs. This follows from their good agreement with theoretical calculations of τ_e, presented for GaAs in paper [29]

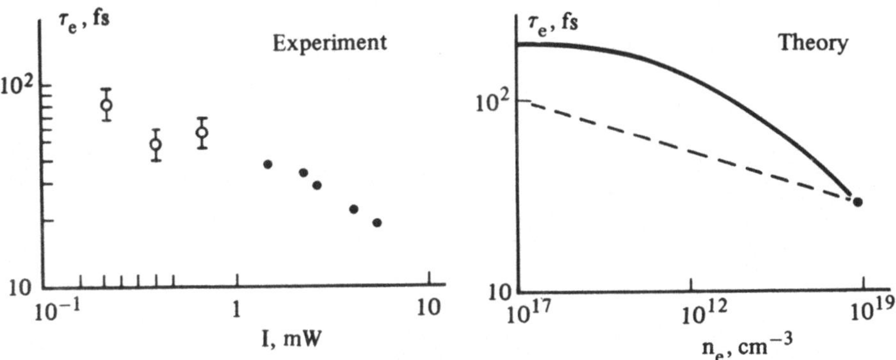

Fig. 3. a) Experimental dependence of the intraband energy relaxation time for electron (τ_e) in $Al_{0.34}Ga_{0.66}As$ on the average intensity of the laser pump I [28] (for I $\sim$ 1 mV, the density of photoexcited electrons $n_e \sim 7 \cdot 10^{18}$cm^{-3}); b) the solid line, – computed in Ref [29], is the theoretical dependence of the time τ_e in GaAs (T $\approx$ 300 K) on the free-carrier density neglecting intervalley scattering; the broken curve is the same, but taking into account intervalley scattering using the data from the experiment.

As we have already seen the simple picture of single-electron intraband relaxation, considered above, is confirmed experimentally. Nevertheless, in the general case it is necessary to take into account the possibility of the collective and nonlinear effects in the dense plasma. For example, electron-phonon interaction screening can result in the increase of time τ_{e-LO} as n_c increases [30]. The critical densities at which the screening influence on τ_{e-LO} becomes substantial, are theoretically estimated to be $10^{19} - 10^{21} \mathrm{cm}^{-3}$, depending on the type of electron-phonon scattering. Experimental evidence supporting the existence of this effect is given in review [30], devoted to the hot carrier investigation in semiconductors with the help of picosecond lasers. Besides, the value τ_{e-LO} can be affected by the change of phonon mode populations owing to phonon emission in the cascade relaxation of carriers to the bottom of conduction band. At very low temperatures (T $\sim$ 4K) and plasma densities considerably smaller than those considered here ($n_c \sim 10^{17} \mathrm{cm}^{-3}$) the exchange and correlation effects are known to radically change the nature of plasma, causing it to condense into an electron-hole liquid [25, 31]. The occurrence of similar effects at high temperatures and high excilation levels is not excluded. The hypothesis of exciton formation and condensation occurring in the conditions of laser annealing was proposed by Van-Vechten (see rewiew [32]). This idea formed the foundation for the so-called plasma model of laser annealing (see Sec. 1.3).

1.2. Electron-phonon relaxation and lattice heating.

Owing to initially large difference of the temperatures of carriers-T_c and lattice-T, the probability of phonon emission by carriers by far exceeds the probability of their absorbtion, so that the rate of energy transfer from hot plasma to the lattice is independent of T. This rate, therefore, can be given as

$$S = \frac{\hbar \omega_0 n_e}{\tau_{e-LO}} \tag{1.3}$$

where $\omega_0 \approx 10^{12} - 10^{13} \mathrm{s}^{-1}$ is the optical phonon frequency. Energy transfer from electron subsystem to nonequilibrium optical phonons occurs over times longer than τ_{e-LO} (see below). At these times hot carrier diffusion changes the value n_c from the estimates presented in Sec. 1.1.
To determine the value of n_e at $t > \tau_{e-LO}$ we shall write the balance equation for the total energy density of the system of hot carriers E_e [33] :

$$\frac{\partial E_e}{\partial t} = G_0(\hbar \omega - E_g) e^{-\gamma z} + <\epsilon_e> \frac{\partial^2 n_e}{\partial z^2} - \hbar \omega_0 n_e \tau_{e-LO}^{-1} . \tag{1.4}$$

18

The first term on the right side describes energy growth owing to photoexcitation of electron-hole pairs; the second term describes the diffusion outflow of energy into the bulk of the medium ($<\epsilon_e> = E_e/n_e$ is the energy per carrier); the last term determines the rate of energy transfer to the lattice.

The density n_e decreases with the distance into the medium over the characteristic distance $l_e = \min(\gamma^{-1}, l_*)$, where l_* is effective diffusion length (see below). From (1.4) one can see that the effective time constant for establishing the stationary state is $\tau_{E_c}^{-1} = Dl_e^{-2} + \tau_{e-LO}^{*-1}$, where $\tau_{e-LO}^* = (<\epsilon_e>/\hbar\omega_0)\tau_{e-LO}$ is the time required for a hot carrier to transfer all its energy $<\epsilon_e>$ to the lattice in the process of $<\epsilon_e>/\hbar\omega_0 \sim 10 - 10^2$ individual acts of phonon emission ($\tau_{e-LO}^* \sim 10^{-12} - 10^{-10}$s, $(D\gamma^2)^{-1} \sim 10^{-11}$s, $D \sim 10$ cm^2/s, $\gamma \sim 10^5$cm^{-1}). At times $t > \tau_e$ we have $\partial E_e/\partial t = 0$ and from (1.4) one obtains the inhomogeneous equation for the stationary density n_e. Its solution has the form ($\partial n_e/\partial z|_{z=0} = 0$):

$$n_e = G_0\,\tau_{e-LO}\,\frac{(\hbar\omega - E_g)\,l_*^{-1}\,\gamma^{-1}}{\hbar\omega_0\,(1 - l_*^{-2}\gamma^{-2})}\left(e^{-z/l_*} - \frac{1}{l_*\gamma}\,e^{-\gamma z}\right). \tag{1.5}$$

Here the effective diffusion length $l_* = (D\tau_{e-LO}^*)^{1/2}$.

At weak optical absorption ($\gamma < l_*^{-1}$) for the rate of energy transfer to the lattice one obtaines from (1.3) and (1.4)

$$S = G_0(\hbar\omega - E_g)\,e^{-\gamma z} \tag{1.6}$$

In this case the laser energy stored in the electron subsystem is tranferred to the lattice at the same rate and in the same volume where it is absorbed.

At strong absorption ($\gamma \gg l_*^{-1}$) we have

$$S = \frac{G_0(\hbar\omega - E_g)\,e^{-z/l_*}}{\gamma l_*} \tag{1.7}$$

Now the absorbed laser energy is tranferred to the lattice in a layer of thinckness l_* determined by diffusion and the rate of energy transfer at the surface $z = 0$ is a factor of γl_* lower than with (1.7). The coefficient of diffusion

$$D = \frac{2kT\,\tau_{e-LO}\,\tau_{h-LO}}{m_e^*\,\tau_{h-LO} + m_h^*\,\tau_{e-LO}}$$

can reach values of $D \sim 10^2$cm^2/s^{-1} and $l_* \sim 10^{-5} - 10^{-4}$cm at high temperatures ($T_c \sim 10^4$K). Hot carrier diffusion therefore can substantially decrease lattice heating rate.

Energy transfer at times τ^*_{e-LO} goes into definite optical lattice vibrations and the probability of acoustical phonon emission by electrons is substantially lower. Acoustical (transverse TA and longitudinal LA) phonons with a non-thermal distribution are generated by the decay of each optical phonon into a pair of acoustical phonons ([42] , Fig. 4).

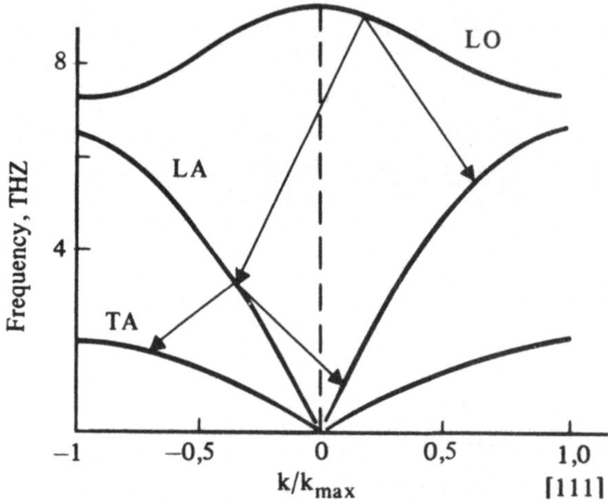

Fig. 4. Schematic diagram of the decay of high-energy long-wavelength longitudinal optical (LO) phonons, produced during the relaxation of hot
carriers, into longitudinal (LA) and transverse (TA) acoustic phonons
with conservation of energy and momentum in each event and an example
of the subsequent chain of decays.

These processes occur over times τ_{LO-TA}, τ_{LO-LA}. Finally, due to the phonon-phonon scattering gradual energy thermalization occurs over times τ_{LA-T}, τ_{TA-T}. It should be noted that because of the difference in the thermalization times ($\tau^{-1} \sim \omega^3_{ph}$) the low-frequency and especially the long-wave length phonons (optical, in particular) can not thermalized at very short times, while the short wave length phonons are already completely thermalized.

The intire process of energy transfer from the initial electron-hole excitation to lattice thermal vibrations can be charecterized by the effective time of energy thermalization τ_{e-T}.

A theoretical estimate of the phonon-phonon relaxation time [34] gives $\tau_{LO,LA-TA} \sim 10^{-12}$s. Experiments on the study of fast phonon-phonon relaxation in semiconductors, carried out at low temperatures (T $\sim$4K) and at low intensities of the exciting radiation ($n_c < 10^{19}$cm^{-3}) [35–38] confirme this estimate.

20

Besides, in a number of experiments performed at the initial temperature $T_{in} = 300$ K and high W (under laser annealing conditions) it was demonstrated that the effective energy thermalization time is very short and is equal to $\tau_{e-T} \lesssim 1$ ps (see review [39] and Sec. 2.2.).

Then, assuming that the energy is thermalized practically instantaneously, one can find the rate of lattice heating by the laser pulse. Using (1.6) we have at the surface (z = 0) neglecting diffusion of heat and hot carriers:

$$c_v \frac{\partial T}{\partial t} = S = G_0(\hbar\omega - E_g) \; ;$$

here c_v — heat capacity per unit volume, T — the lattice temperature. For $\omega \sim 4 \cdot 10^{15} c^{-1}$, $\tau_p \sim 30$ ns, $W \sim 1 J/cm^2$, $c_v \sim 1 J/cm^3 K$ we have $\partial T/\partial t \sim 10^{12} deg/s$. Even higher values of $\partial T/\partial t$ are reached by using picosecond pulses ($G_0 \sim 10^{33} cm^{-3} c^{-1}$). These very high rates of pulsed laser heating, which cannot be achived by ordinary thermal heating of semiconductor surfaces, provide unique possibilities for stimulating nonequilibrium phase transitions at the surfaces. Ultrafast linear and nonlinear laser spectroscopy techniques are most adequate for the study of their kinetics (see Sec. 1.3, 2.2, 2.3).

A more accurate quantitative description of lattice heating by a laser pulse under the condition $\tau_p > \tau_{e-T}$ is obtained thermodynamically with the help of the solution the heat conduction equation. In this equation the heat source S is determined by the rate of energy transfer to the lattice under the condition that the energy is instantaneously thermalized.

The heat conduction equation is the starting point of the investigation of the laser annealing process from the view point of the thermal model [39—41] (see Sec. 1.3, 2.3.1). Equations of this type are also used in the analysis of a number of nonlinear-optical processes which occur on the surface of semiconductors, metals, dielectrics and their melts due to spatially and temporally periodic lattice heating by laser radiation (see Sec. 3,4).

1.3. Pulsed laser annealing of a semiconductor surface. Thermal and plasma annealing mechanisms

The phenomenon of pulsed laser annealing (PLA) consists in extremely rapid (usually within several tens of nonosecond) restoration of the crystalline structure of previously disordered or even completely amorphouse semiconductor surface layer under the action of rather high power laser pulse with the energy quantum exceeding the forbidden band width.

This interesting and practically important effect discovered in USSR in mid 1970-s [5, 6] has now been thouroughly investigated and found wide applications in semiconductor materials technology (see reviews and books [6–12, 301–304] PLA is most widely used to eliminate structural imperfections and radiation defects introduced into the crystal surface layer by ion implantation, i.e. by doping proper impurities into this layer through the irradiation of the surface by the accelerated ions of these impurities [43].

PLA, for example, provides for the thechnology an important possibility of obtaining perfect crystalline structures in surface layers with dopant concentrations, unattainable with the conventional thermal annealing (up to 10^{21}cm^{-3} and higher [44]).

Obtaining such great impurity concentrations by conventional annealing of structural defects, induced by ion bombardment, due to slow heating at 900–1000°C [43] in the furnace is hampered by impurity particle diffusion into the bulk of the material [45]. In the process of the ultrafast PLA the length of diffusion is strongly reduced.

Aside from applications in microelectronics, the perfect crystalline layers with ultrahigh concentrations of minority carriers obtained by means of PLA are also interesting from the physical point of view. Under these conditions it is possible to observe processes occuring because of the presence of a dense plasma (phonon mode softening with increasing carrier concentration [46], and change in the forbidden band width depending on n_c [47]), as well as to record the optical and electrical phenomena in the semiconductors depending on n_c [48].

The spatial coherence of laser radiation used for PLA, allows to produce periodic spatial structures at the crystal surface being annealed, for example, gratings formed by alternative crystalline and amorphous regions (so called, interference laser annealling [49, 50]).

Ultrafast recovery of a perfect crystalline structure in PLA was proposed to be used for narrowing of the gamma emission line of short-lived nuclear isomers excited by pulsed neutron pumping [51]. High power neutron pumping required for obtaining gamma-radiation enhancement is certain to destroy a crystalline structure and eliminate the possibility of utilizing the Mössbauer and Bormann effects for the optimization of the parameters of a gamma laser. The use of PLA for the crystalline structure recovery within several tens of nanoseconds after the excitation of nuclei can help solve this problem [51]. Since the discovery of laser pulsed annealing, great theoretical and experimental efforts have been made for classification of the physical mechanism of PLA. Thus far in the main two alternative mechanism of PLA — "thermal" and "plasma" ones were considered.

The conventional "thermal" model of pulsed laser annealing [39–41] is based on the concept of fast energy transfer from the hot carrier subsystem to the lattice (see Sec. 1.2). According to this model, when the laser pulse energy density W is sufficiently high, the amorphized layer 50–500 nm thick melts, undergoing phase transition of the first order. The melt front moves rapidly from the surface into the bulk of the material and reaches the crystalline substrate. When the melt front moves back to the surface during the cooling of the sample after the end of the pulse, epitaxial crystal growth occurs. Analytical estimates, based on the assumption of fast laser pulse energy transfer to the lattice (see Sec. 3.8) and numerical solution of heat-conduction equation show that melting temperature is easily reached at values of W typical for PLA (see also [44, 46]).

According to Lindemann's criterion, the crystal melting ($T = T_m$) starts when the mean-square displacement of an atom from its equilibrium position ($<U^2>$) is a definite fraction x of the squared unit-cell size a^2 (for most materials $0.2 < x < 0.25$). At sufficiently high temperatures

$$<U^2> = \frac{9\hbar^2 T}{M k T_D^2} = \frac{1}{MN} \sum_j <|q_j^2|> \qquad (1.8)$$

where M is the atomic mass, T_D is Debye temperature ($T > T_D$), N is the number of atoms in the crystal, q_j is the amplitude of jth normal mode of the acoustical vibrations of the crystal.

Then the melting temperature is equal to

$$T_m = \frac{x}{9\hbar^2} M k T_D^2 a^2 \; .$$

(For Si, for example, $T_D = 625$ K, $a = 1{,}18$ A, $x = 0{,}2$ and from given formula we have $T_m = 1685$ K in agreement with the experimental value).

It is not yet clear whether the solid-liquid phase transition occurs after the energy is thermalized over all phonon modes j ($1 < j < N$) entering into eq. (1.8), or when only several of the most highly pumped phonon modes have such large amplitudes q_j that Lindemann criterion starts to be met.

The thermal model of laser annealing is confirmed by the results of comprehensive studies of semiconductor surface during and after termination of the laser pulse. This includes great number of experiments on ultrafast laser spectroscopy of strongly excited semiconductor surface (see Sec. 2). The measurements of velocities of atoms evaporated from the surface in laser annealing show that the surface temperature reaches approximately 2000 K, well above the melting temperature of Si. The thermal model is also confirmed by measurements of temporal evolution of the temperature with the help of syncrotron X-ray radiation [54, 55], the photoemission and electrical conductivity [56, 57] and impurity distribution studies after laser annealing [58, 44, 45].

A number of investigators, however, defend another pulsed laser annealing mechanism, the so called plasma or collective PLA model [32, 59, 304].

While in the thermal model the melting of a crystal occurs due to the intensive thermal motion of lattice atoms, the solidliquid phase transition in the plasma model is caused by the softening of transverse acoustical phonons in covalent semiconductors, accompaning electron-hole plasma concentration growth:

$$\omega_{TA} = \omega_{TA}^0 \left(1 - \frac{f \epsilon_\infty n_c}{4n} \right) ,$$

where $f \sim 1$ (for Si $f = 0.85$), ϵ_∞ is the dielectric constant of the crystal, n is the atomic density. This phenomenon is determined by the fact that in the process of exitation from the valence band to the conduction band an electron makes the transition from a bonding state into antibonding state so the covalent bond is weakend. The acoustical phonon softening can also lower the temperature of the "ordinary" melting [60].

The plasma model is based experimentally on the results of the lattice temperature measurements by the method of Raman scattering (RS) of light (see Sec. 2). The temperature obtained by measuring the ratio of the intensities of Stokes and anti-Stokes components of RS in Si during laser pulse annealing in a number of experiments turned out to be equal to only 600°K [61, 62]. The results reported in the first papers on Raman scattering were criticized by a number of authors [63, 64], but in subsequent publications on the subject the correctness of the first results was advocated by Compaan et al [65, 66]. The work [212] by von der Linde's group seems to have put an end to this discussion, Compaan et al have agreed on it [217].

Another difference between the predictions of the two models which can be tested experimentally is the difference in the symmetry of the phase formed by the laser pulse. In the thermal model the resulting melt is an isotropic medium, whereas in the plasma model, according to [67], dipole-dipole interactions between excitons should lead to the formation of noncentrosymmetric exciton state with the T_d symmetry (similar to the symmetry of GaP and GaAs crystals). The second harmonic generation in reflection is sensitive to the surface layer symmetry. The first experiments on the second harmonic generation of probing radiation in a reflection in GaAs [18] give evidence in support of the thermal model (see Sec. 2.3). The same conclusion was made later in work [16], where the second harmonic of probe radiation from Si was observed, when Si was acted upon by high-power femtosecond laser pulse.

Thus, although most experimental data on the study of laser pulse annealing support the thermal model, there are some experiments contradicting it and which are consistent with the plasma model [32, 304].

Another nonthermal model of laser annealing was proposed in [70]. According to this model the Paierls dielectric-to-metal and metal-to-dielectric phase transition occurs in annealing. In the frame of this model the possibility of defects annealing due to the excitation of the relative vibrations of crystal sublattices was considered [71]. It was not until recently that the model of cool lattice melting due to static acoustical displacement instability induced by the laser pulse was developed [190, 307].

One may, therefore, conclude that the lack of a complete physical picture of the laser radiation interaction with the semiconductor surface is, to a great extent, due to the lack of experimental diagnostics methods adequate to the problem under study. It is for this reason that new methods of diagnostics of laser-induced phase transitions on the surfaces of solids have been recently developed in a number of laboratories. These methods will be dealt with in the next section (see also the conclusions).

2. LINEAR AND NONLINEAR OPTICAL METHODS FOR RECORDING THE STATES AND LASER-INDUCED PHASE TRANSITIONS ON THE SURFACE OF CONDENSED MEDIA

As has already been noted in the Introduction, the most dramatic result of laser technological advances in the last several years is the development and application of lasers emitting ultrashort pulses of femtosecond duration [13—15, 72—75]. Of course, we cannot give an exhaustive review of the application of ultrashort laser pulses to the study of all types of the fast photoprocesses in solids (there already exist a good review of the early works on this subject [30], collective monographs [74, 76] and proceedings of international conferences on ultrafast phenomena [14, 75, 77—80]). Nevertheless, in this section we will give a brief description of recent studies aimed at the application of pulsed laser techniques for diagnostics of fast processes on the surfaces of solids (Sec. 2.1); we will study in greater detail the nonlinear optical aspect of these studies in Sec. 2.2, 2.3.

2.1. New data from linear optical spectroscopy on fast laser-induced processes on semiconductor and metal surfaces

2.1.1. The study of the optical reflection from surfaces in real time in pulsed laser annealing regime

The works of D.Auston and his collaborators on determining the time dependence of linear optical reflection at the semiconductor surface subjected to PLA [81, 82] served as an impetus for the wide application of nonperturbing local, highly imformative optical spectroscopy methods for the diagnostics of fast laser-induced phase transitions on the surfaces of solids. In the course of these studies the above mentioned controversy, which was later resolved, regarding the mechanism of pulsed laser annealing arose and important data on the processes, occuring on semiconductor and metal surfaces were obtained. In a paper [16] by Shank and his colleagues the femtosecond dynamics of optical reflectivity change on the Si surface subjected to PLA under the action of 90 fs pulse was studied for the first time. The main laser was mode-locked dye laser, in which the technique of colliding pulses was used. A single 90 fs pulse, piked out by the electro-optical shutter, was amplified in the cascade of pulsed amplifiers and divided into two channels. The pulse in one channel was used for annealing, the weaker pulse in the other channel was probing one. To carry out measurements in a wide spectral range the probe pulse frequency was converted into a broad continuum, covering practically all visible and including the near infrared ranges by stumulated Raman scattering and accompanying nonlinear processes in a cell with heavy water. The process of carrying out the measurements and collecting experimental data was controlled by a microcomputer. The probe pulse reflection from crystalline Si (111) surface in various spectral regins as a function of the delay of the probe pulse relative to the annealing pulse was studied.

Fig. 5 gives experimental data [16] for three values of the probing wavelength: λ_p = 1000, 678 and 440 nm. The annealing pulse energy density W is given in the units of threshold energy, W_{thrsh} = 0.1 J/cm^2, the latter was determined by the appearance of amorhous films on the Si surface.

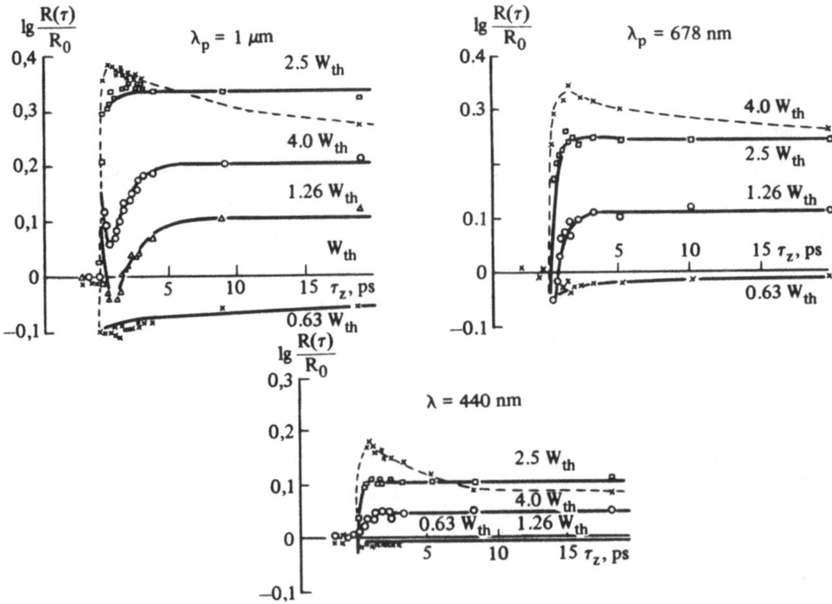

Fig. 5. The reflection of the (111) surface of crystalline silicon as a function of the delay time between the annealing (λ = 620 nm, τ_p = 90 fs) and probing pulses for three values of the probing wavelength [16]. The solid lines with W = 0.63 W_{th} were calculated taking into account carrier diffusion inside the crystal; the solid lines for W > W_{th} were calculated on the basis of the model of a thin layer of melt of variable thickness; the broken curves with W = 4 W_{th} were drawn "freehand".

The authors of [16] proposed a rather simple physical model, accounting for the variety of the obtained data.

According to this model, at the earliest stage, several hundreds of femtoseconds after the annealing pulse is absorbed in the surface layer with a thickness of

$d \approx \gamma^{-1} \approx 3$ μm (γ is coefficient of absorption of the annealing radiation with $\lambda_p = 620$ nm), the changes in the probe pulse reflectivity with $\lambda_p = 1000$ and 678 nm are caused by a dense photestimulated electron-hole plasma (previously, this stage was also observed by Von der Linde et al. [83] and Bloembergen et al. [84]).

The optical properties of the plasma are discribed by the dielectric constant:

$$\epsilon(\omega) = \epsilon_0 \left(1 - \frac{\omega_p^2}{\omega^2}\right), \qquad (2.1)$$

where $\omega_p = (4\pi n_e e^2/m^*\epsilon_0)^{1/2}$ is the plasma frequency, n_e is the carrier density, m^* is the effective mass, ϵ_0 — characterizes lattice and interband transition contribution to the dielectric constant.

The estimates show that for $W \approx 0,6_{th}$ one obtains $n_e \approx 5 \cdot 10^{21}$ cm^{-3} for $m^* \approx m_e$ — the free electron mass (however, see criticism of this approximation in [85]). As W increases, both n_e and ω_p increase, so that the drop in the coefficient reflection at $\lambda = 1$ μm, when $\omega_p > \omega$ is replaced by the rise, when $\omega_p < \omega$. Comparing Fig. 5 a–c, one can conclude, that the maximum value of ω_p corresponded to the wavelength λ_p, 678 nm $< \lambda_p <$ 1000 nm.

The change in the coefficient of reflection at long time delays is explained by heating and fast melting of the surface layer owing to the energy deposition into the lattice from plasma and to the melting front advance into the crystal.

The solid lines in Fig. 5 were calculated by taking into account these two stages. In this experiment, therefore, the optical characteristics of highly excited Si (in which about 10% of its valence electrons are excited into the conduction band) were recorded for the first time before and during the subpicosecond melting of its crystalline lattice. The melting process proceded on time scales (hundreds of femtosecond) comparable to the period of the vibrations in LO-phonon (70 fs). This fact forces the authors of [16] to raise an important conceptual question of what is meant by "melting" during such short time intervals. Of course, the question formulated in Sec. 1 about the presence (or absence) of thermodynamic equilibrium in the phonon subsystem in such short time intervals also remains open.

It should be noted, however, that the results presented in [16] as well as Auston's earlier results [81, 82] and other linear optical spectroscopy data can be also interpreted from the point of view of the plasma model of PLA [32] (see Sec. 1.1). To get an unambiguous solution of the problem requires the application of nonlinear optical spectroscopy data carrying structural information (see Sec. 2.3).

A great amount of experimental data on the picosecond stage of PLA of Si and Ge surfaces are presented in the works of Bloembergen et al [39, 84, 86–90]. These authors have also used the method of optical reflectivity change measurements of the surface being annealed.

Together with optical reflection, diagnostics of the PLA process is also performed by measuring the temporal behavior of the optical transmission of thin semiconductor films under pulsed laser action. The interpretation of the transmission data evidently requires a knowledge of the distribution of the absorption coefficient both along the direction of propagation of the probing beam and across the laser beam. It is clear that these data are not always accessible, so that in interpreting the transmission data, errors can arise. Thus, the assertion made in [91] on the basis of Si thin filsm transmission data about the absence of melting in PLA turned out to be incorrect (see [92–94]). Very promising are the time-resolved experiments with x-ray difraction [245, 246]. There are some unexplained peculiarities in the time-resolved optical reflection studies in Si [260].

2.1.2. Raman and stimulated Brillouin scattering of light by the laser-excited surfaces of semiconductors

The results of temperature measurements of the Si surface subjected to PLA with the help of spontaneous Raman scattering (RS) of light performed by Compaan and his coworkers gave rise to substantial disagreements and a long standing controversy. In these experiments the ratio of the intensities of the Stokes I_s and the anti-Stokes I_{as} RS components by the 522 cm^{-1} mode of the crystalline Si was determined, this ratio being the measure of heating of the substance under investigation. The lattice temperature can be calculated from the following equation [61, 63]:

$$\frac{I_s}{I_{as}} = C\,\frac{n+1}{n} = C\exp\left(\frac{\hbar\Omega}{kT}\right)\,, \tag{2.2}$$

where n is the occupation number of the phonon modes, Ω is the frequency the phonon studied, T is the lattice temperature;

$$C = \left(\frac{\omega_s}{\omega_{as}}\right)^3\,\frac{\tau_s}{\tau_{as}}\,\frac{\gamma+\gamma_{as}}{\gamma+\gamma_s}\,,$$
$$\omega_s = \omega - \Omega\,, \quad \omega_{as} = \omega + \Omega \tag{2.3}$$

29

are the frequencies of Stokes and anti-Stikes Raman scattering components respectively, $\tau_s(\tau_{as})$ are the Stokes (anti-Stokes) RS cross-section, γ, $\gamma_{s,as}$ are the absorption coefficients at the frequencies ω, ω_s, ω_{as} respectively.

The first lattice temperature measurements using this technique [95, 96] gave the value $T < 700$ K, that is considerably lower than the melting temperature of Si ($T_m \cong 1680$ K) for $W \approx 0.3 - 0.4$ J/cm^2, which obviously exceeds the threshold value typical for PLA. Later these results were questioned by a number of authors [63, 64, 87, 97] and a repeat experiment by Von der Linde and Wartmann [62, 63] initially gave a much higher value of T, equal to or exceeding the value of T_m for silicon. Subsequent more accurate measurements by Compaan's group [61, 65] and afterwards a joint experiment performed by Compaan and Von der Linde [66] gave results which were close to the initial data, obtained by Compaan's group, i.e. showed the Si sample surface temperature to be twice lower than the melting temperature under normal conditions. Later it was, however, acknowledged that with taking into account all the sources of errors, one can obtain the surface temperature during PLA, which is close to the melting temperature [212, 217].

The main difficulties of the experiments on determining T from RS spectra are connected with the high spatial and temporal resolution required in the experiment, and also with the difficulties of determination the factor C in equation (2.2) and with the temperature dependence of the position of the RS lines.

As has already been noted in Sec. 1, most experimental data, including direct determination of the surface temperature from measurements of the velocities of evaporated atoms [53] and x-ray scattering [54, 55] unambiguously show, that silicon surface is melted in PLA. It should be noted, however, that Raman scattering is due to phonons from the center of the Brillouin zone with practically zero wave vectors whereas melting occures mainly due to shortwave length phonons from the periphery of the first Brillouin zone. The latter phonons can be "hot" enough for Lindemann criterion (1.8) to hold whereas the former remain relatively cold owing to the supposed phonon-subsystem nonequilibrium (see Sec. 1.1). To clarify the situation, data from coherent active Raman spectroscopy with high temporal and spectral resolution could be very useful (see [21, 214] and Sec. 2.2.5).

When surface layers of semiconductor crystal are strongly excited owing to direct interband absorption of laser radiation, the signals of Brillouin scattering also exhibit anomalies in some cases. In paper [98] a marked increase of Stokes signal of Brilluoin scattering on surface acoustic waves in GaAs was observed at some threshold value of intensity of argon laser radiation used to excite these spectra. The observed growth was by far greater to be attributed only to the lattice temperature increase caused by the increase of the absorbed energy of optical radiation. The effect can be related to the stimulated acoustic phonons generation under interband electronic transitions [99, 307].

2.1.3. Optical recording of periodic surface ripples

Interference pulsed laser annealing (IPLA) [49, 100] realizable with two crossed coherent light beams, results in the formation of periodic sequence of crystalline (annealed) and amorphous (unannealed) areas in the surface layer of the semiconductor. This artificial periodc surface structure, easily observed by the diffraction of a probing optical beam incident on it, is a convenient tool for the study of the conditions of epitaxial growth of surface crystalline layers in PLA. Such experiments allow to determine quite accurately the energy density threshold and the depth of the crystallization layer. The formation of such a structure and its basic characteristics can be easily described on the basis of the ordinary PLA thermal model [101].

At the same time by changing the spacing of the interference pattern formed by the annealing beams and laser radiation parameters one can easily control the conditions of PLA, in particular, it is possible to create considerable temperature gradients along the surface [102, 103].

This technique enabled Alferov et al. [103] to discover, aside from the main "amorphization ring", an additional amorphous ring around the GaAs area, irradiated by picosecond laser pulse, and to attribute it to a new, so far unknown, GaAs crystalline phase.

The observation of a diffraction pattern from the probing beam in the far field on the periodically modulated surface relief has recently become the standard method for recording such perturbations, which are spontaneously generated at the surface of solids under the action of high power laser radiation (for the first time this the method was used for these purposes in [104, 190] (see Sec. 3).

2.2 Nonlinear optical diagnostics of laser-indiced transitions on a semiconductor surface; generation of optical harmonics and combination frequencies in reflection

2.2.1. Nonlinear optical diagnostics of the surface

Substantially new data on the state of the surface layers can be obtained from the analysis of their nonlinear response, observed in nonlinear reflection, i.e., in the appearance of optical harmonics and combination frequencies in the reflected light beam. The physics of the phenomenon is quite simple (see below, Fig. 6).

Optical harmonics and combination frequencies, which are absent in the spectrum of light, incident on the interface, appear in the reflected light due to the uncompensated "backward" radiation of optically induced oscillators from the boundary layers of nonlinear medium. Because of the nonlinearity of the oscillators in the reflecting medium the "backwards" reradiated field, which in classical optics gives rise to the usual "Fresnel" reflection, contains, in addition to the frequency components initially present in the spectrum of the incident light, also new spectral components with multiple and combination frequencies.

It is to be mentioned that the mechanism of the nonlinear reflection, described above, was discovered more than 20 years ago [105].

In [18] we have drawn attention to the fact, that anisotropy and nonlocal nature of the nonlinear response of the surface layer (in many cases they are significant when linear response is isotropic and local) appear to be the sources of unique information about fine details of the crystalline structure. Subsequent work showed nonlinear optical methods to be extremely informative and fast and to have high spatial resolution.

The application of modern means for filtering and accumulating weak optical signals as well as the use of intensive pico- and femtosecond pulses following one after another with high frequency, now enable easily and reliably recording the nonlinear optical surface response and using it to obtain structural information about the surface under study (which is often inaccessible to linear optical methods).

At present these effects are used to observe molecular layers, absorbed on surfaces of the solid, surface excitations and their interactions with one another and with bulk electromagnetic waves [106—112].

In the works [18, 113, 114], carried out in our laboratory, the phenomenon of second harmonic generation (SHG) in the reflection from the surface of noncentro-symmetrical GaAs crystal (class 43 m) was first used to study the PLA dynamics. The essence of the proposed technique can be described in the following way: the surface layer of the crystal becomes centrosymmetrical as a result of melting (or amorphization due to ion implantation) and does not make dipole contribution to the reflected second harmonic. The recovery of the crystalline lattice in the process of PLA causes the appearance of the reflected wave of probe radiation second harmonic. The measurement of SH intensity variation in time allows to study the dynamics of sample surface recrystallization during PLA. The analysis of second harmonic polarization characteristics yields information on the quality of amorphous and restored crystalline lattice; while measurement of the SHG time dependence gives information about the dynamics of phase transitions at crystal surface during PLA.

Soon after the publication of our first results on SHG in GaAs [18] (see, also, note in [128]), studies in which SHG was reported to have been observed in the reflection from centrosymmetrical Si crystal under the conditions of strong excitation close to PLA appeared. The effect appears to have been incorrectly interpreted in [130] in terms of crystal symmetry change and Frenkel-exciton generation (see, also [218]). Having correctly interpreted SHG in Si as the manifestation of nonlocal nonlinearity of quadrupole type (see Sec. 2.3.4), Shank et al. [126] applied this technique for the diagnostics of structural surface changes in the first stage of PLA, that of surface layer melting- on a subpico-second time scale. Heinz et al [154, 219] have used the anisotropy of nonlinear response of the "reconstructed" silicon surface to investigate phase transitions at a surface.

2.2.2. Theory of generation of optical harmonics in reflection. Nonlinear optical source

The starting point for the theoretical analysis of nonlinear reflection are Maxwell's equations with the corresponding nonlinear sources and boudary conditions, taking into account the continuity of the tangential components of the fields and normal components of the displacement currents at the interface [115–118, 220, 221].

We can, by using the definitions of Ref. [118], write down for the components of the bulk nonlinear polarization vector P^{NL} the following expression:

$$P_i^{NL} = N \left(<d_i>_{av} - \nabla_j <Q_{ij}>_{av} \right),\qquad (2.4)$$

where

$$d_i = -ex_i, \quad Q_{ij} = -\frac{1}{2}\, ex_i\, x_j$$

are, respectively, the components of the dipole vector and the quadrupole-moment tensor, arising in each unit cell in the volume of the crystall under the action of the optical field; the angular brackets $<...>_{av}$ indicate averaging over all electrons in the cell; N is the total number of unit cells per unit volume; the gradient operator, ∇_j, operates along coordinate x_j, which gives the position of the selected cell; $x_{i,j}$ characterizes the part of the relative displacement of the electrons, which is nonlinear in the field within one cell. Here and below repeated indices are summed over from 1 to 3.

The expansion of the source in powers of the field of the incident wave has the form

$$P_i^{NL} = P_i^{(2)} + P_i^{(3)} + \ldots , \qquad (2.5)$$

$$P_i^{(2)} = \chi_{ijk}^{(2)D} E_j E_k + \frac{1}{2} \chi_{ijkl}^{(2)D} \nabla_j E_k E_l , \qquad (2.6)$$

$$P_i^{(3)} = \chi_{ijkl}^{(3)D} E_j E_k E_l + \frac{1}{3} \chi_{ijklm}^{(3)Q} \nabla_j E_k E_l E_m ; \qquad (2.7)$$

here E_j and E_k are the components of the electric field vector of the incident light wave; $\chi_{ijk}^{(2)D}$, $\chi_{ijkl}^{(3)D}$ are the tensors of the dipole and $\chi_{ijkl}^{(2)Q}$, $\chi_{ijklm}^{(3)Q}$ are the tensors of the quadrupole quadratic and cubic susceptibilities, respectively:

$$N <d_i>_{av} = \chi_{ijk}^{(2)D} E_j E_k + \chi_{ijkl}^{(3)D} E_j E_k E_l + \ldots ,$$

$$N <Q_{ij}> = -\frac{1}{2} (\chi_{ijkl}^{(2)Q} E_k E_l + \chi_{ijklm}^{(3)Q} E_k E_l E_m + \ldots) . \qquad (2.8)$$

The tensors $\chi_{ijk}^{(2)D}$, $\chi_{ijkl}^{(2)Q}$, $\chi_{ijkl}^{(3)D}$, ... describe the properties of the medium giving rise to the corresponding "bulk" nonlinear optical response. According to Neumann's crystallographic principle, the symmetry of these tensors and, therefore, the number and mutual coupling of their nonzero components, are determined by the point symmetry group of the crystal. The general symmetry rules are the same for tensors with the same rank, so that, for example, the crystallographic "selection rules" for the components of the tensors $\chi_{ijkl}^{(2)Q}$ and $\chi_{ijkl}^{(3)D}$ are the same. And, vice versa, the information about the crystalline structure which can be obtained by determining the symmetry of the last two tensors is, generally speaking, the same.

In what follows we shall require data on the structure of the nonlinear susceptibility tensors of a number of cubic crystals, belonging to the centrosymmetric class m3m (to which crystalline silicon and germanium belong) and the non-centrosymmetric class $\bar{4}$3m (to which gallium arsenide and gallium phosphide as well as some other crystals belong). According to Refs. [119] and [120] we have:

a) the class m3m: 1) all $\chi_{ijk}^{(2)D} = 0$. 2) Twenty-one of the 81 components of the tensors of rank 4 $\chi_{ijkl}^{(3)Q}$ and $\chi_{ijkl}^{(3)D}$ differ from zero, and of these only four are independent. In terms of the principal crystallographic axes these components have the form

34

$$\chi_{1111} = \chi_{2222} = \chi_{3333}, \quad \chi_{1122} = \chi_{iijj}, \quad \chi_{1212} = \chi_{ijij}, \quad \chi_{1221} = \chi_{ijji} \qquad (2.9)$$

where $i \neq j$ and $i,j = 1, 2, 3$. The quantity $\zeta_{Q,D}$

$$\zeta_{Q,D} = \chi_{1111}^{Q,D} - (\chi_{1122}^{Q,D} + \chi_{1212}^{Q,D} + \chi_{1221}^{Q,D}) \qquad (2.10)$$

is the parameter of the nonlinear optical anisotropy (in linear optics these crystals are isotropic), since in an isotropic medium this quantity vanishes for both dipole and quadrupole nonlinearitie.

b) The class $\overline{4}3m$: 1) The dipole quadratic susceptibility has one independent component

$$\chi_{123}^{(2)D} \neq 0 \qquad (2.11)$$

and five more components, obtained by permuting the indices, equal to it. 2) The tensors of rank four have the same structure as in the class m3m.

In the expansion (2.6) and (2.7) the quadrupole terms, as a rule, are much smaller than the dipole terms. An important exception are crystals with a center of inversion and isotropic media, where all $\chi_{ijk}^{(2)D} = 0$ and for this reason the effects which are quadratic in the field, for example SHG, are described by the quadrupole as well as dipole surface (see below) terms.

It is clear that the symmetry of the surface of a crystal differs from that of the "bulk" part of the crystal (even without taking into account the unavoidable "reconstructions" and the appearance of adsorbed layer), so that "surface" terms with an analogous structure must, generally speaking, be added to the expansion (2.6). We shall assume that these terms correspond to the contribution of a thin (several atomic layers thick) nonreconstructed surface section of the crystal to nonlinear polarization. The symmetry group of this layer is evidently a subgroup of the point symmetry group of the entire crystal.

Surface optical nonlinearity at an interface between two media can be characterized by an effective "local" surface nonlinear susceptibility that includes both local (dipole) and nonlocal (quadrupole and magnetic-dipole) responces of the interface layer to the field [221]. Structural discontinuity of the matter in the normal direction to the interface as well as normal to the interface field components discontinuity turn out to contribute both to the surface nonlinearity, giving rise to local and nonlocal contributions into the nonlinear source, respectively. However, with specially defined surface nonlinear susceptibility $\chi^{(2)S}$ one can fully characterizes the overall second-order polarization of the interface layer [221]:

$$P_i^{(2)S} = \chi_{ijk}^{(2)S} \, \epsilon_j(z=0) \, \epsilon_k(z=0)$$

Here $z = 0$ gives the plane of the interface, x, y axes lye in the interface surface; $\epsilon_x = E_x$, $\epsilon_y = E_y$ and $\epsilon_z = D_z$, where D is the electric displacement vector. Note, that we have purposedly defined the surface nonlinear polarization in terms of field components continuous across the interface, i.e., the tangential components of the electric field E and the normal component of the displacement vector D in order to avoid confusion raised by the field variation actoss the interface layer.

Taking into account this circumstance the expression for the full polarization which is quadratic in the field assumes the following form:

$$P_i^{(2)} = \chi_{ijk}^{(2)D} E_j E_k + \chi_{ijk}^{(2)S} \epsilon_j \epsilon_k + \frac{1}{2} \chi_{ijkl}^{(2)Q} \nabla_j E_k E_l + \chi_{ijkl}^{(3)D} E_j E_k E_l + \ldots \qquad (2.6)$$

Here the susceptibilities marked with the index S describe the surface contribution, and their symmetry is fixed by the surface structure. In most cases this contribution can be neglected compared to the bulk contribution, but in the particular case of a centrosymmetric medium this cannot be done, because the dipole surface and quadrupole bulk contributions to the quadratic polarization may turn out to be comparable (for example, in silicon — see below and Ref. [19, 220, 221]) while the dipole bulk contribution, as already pointed out, vanishes.

There is a total of 18 possible independent elements of the surface susceptibility tensor $\chi_{ijk}^{(2)DS}$. The actual number of independent nonvanishing elements, however, can be drastically reduced by the symmetry of the surface or interface layer.

For instance, in centrosymmetrical crystals of class m3m crystallographic face (111) belongs to the symmetry group 3m, and the nonzero components of $\chi_{ijk}^{(3)DS}$ are as follows (all components are represented only by its indices);

$$yyy=-yzz=-xxy=-xyx; \quad xxz=xzx=yyz=yzy, \qquad zxx=zyy; zzz . \qquad (2.12)$$

Here the z axis is along the normal to the surface, i.e. along the [111] crystallographic direction; y axis lies in the mirror plane at the surface (along [211]); x axis is perpindicular to the first two (i.e., along [011]).

In the same crystal, the face (011) belongs to 4m symmetry group; here we have:

$$xxz=xzx=yyz=yzy; \quad zxx=zyy; xyz=xzy=-yxz=-yzx; zzz . \qquad (2.12')$$

Here x,y,z — axes coincide with the principle coordinate axes of the crystal.

36

In the case of the surface of a centrosymmetric isotropic medium the nonzero components of the quadratic surface nonlinear susceptibility are as follows:

$$xxz=xzx=yzy; \quad zxx=zyy; \quad zzz . \tag{2.12''}$$

(z — axis again is directed along the normal to the surface).

In concluding this section, we shall present expressions which are more compact than (2.6) and (2.7) and which will be used below for the quadratic polarization $P^{(2)}$ (2ω) at the frequency of the second harmonic (SH) of the incident monochromatic field $E = (1/2)E_0 e^{-i\omega t} +$ c.c. and for the cubic volume dipole polarization $P_i^{(3)D}(3\omega)$ at the frequency of the third harmonic (TH) of the same field in cubic crystals and isotropic media. From the definitions (2.6) and (2.7) and using (2.8) and (2.10), we obtain the following expression in the principal crystallographic system of coordinates of centrosymmetrical cubic crystals m3m for the quadrupole bulk polarization:

$$P_i^{(2)Q}(2\omega) = \frac{1}{2} \chi_{ijkl}^{(2)Q} \nabla_j E_k E_l = \chi_{ijkl}^{(2)Q} E_j \nabla_k \nabla E_l = \zeta_Q E_i \nabla_i E_i + \chi_{1122}^{(2)Q} E_i (\nabla E) +$$

$$+ \frac{1}{2} \chi_{1212}^{(2)Q} \nabla_i (EE) + \chi_{1221}^{(2)Q} (E \nabla) E_i \tag{2.13}$$

In (2.13) the second and fourth index in $\chi_{ijkl}^{(2)D}$ correspond to the frequency ω of the optical field, while the third index corresponds to the operation of taking gradient. For dipolar cubic polarization at the frequency of the TH in cubic classes m3m and $\overline{4}$3m we have:

$$P_i^{(3)D}(3\omega) = \zeta_D E_i^3 + (\chi_{1122}^{(3)D} + \chi_{1212}^{(3)D} + \chi_{1221}^{(3)D}) E_i (EE) \tag{2.14}$$

In (2.13) and (2.14) summation over the repeated index i is not performed. The expressions (2.13) and (2.14) are also valid for isotropic media, if we set $\zeta_{Q,D} = 0$.

Instead of the quadrupole susceptibilities $\chi_{ijkl}^{(2)Q}$ in isotropic media and cubic crystals a set of different parameters β, γ and δ is sometimes introduced [19, 116]:

$$\beta = \chi_{1122}^{(2)Q}, \quad \gamma = \frac{1}{2}\chi_{1212}^{(2)Q}, \quad \delta - \beta - 2\gamma = \chi_{1221}^{(2)Q} \tag{2.15}$$

Evidently, in an isotropic medium $\delta = \chi_{1111}^{(2)Q}$. In addition, in an isotropic nonconducting medium, as Bloembergen et al. showed, [116] in the low-frequency limit (i.e., for $\omega \ll E_g/\hbar$, where E_g is the gap width) the following relations hold:

$$\beta = -2\gamma = \frac{3}{4N_e}(\chi^{(1)D}(\omega))^2, \quad \delta = 0, \tag{2.16}$$

where $\chi^{(1)D}(\omega)$ is the linear (dipole) susceptibility of the medium and N is the density of the valence electrons.

2.2.3. Field of reflected second and third harmonics generated by nonlinear source

Let a plane wave of frequency ω be incident from the vacuum on the flat boundary of the solid at the angle θ. We choose the coordinate system so that the boundary coincides with the plane $z = 0$, the plane of incidence with the plane $y = 0$; the substance occupies half-space $z \leqslant 0$. Then the incident light wave is described by the vector

$$\mathbf{E} = \frac{1}{2}\mathbf{e}_i\, E\exp(-\frac{i\omega}{c}z\cos\theta - \frac{i\omega}{c}x\sin\theta)\exp(-i\omega t) + c.c. =$$

$$= \frac{1}{2}\mathbf{e}_i\, E\exp(ik_z z + ik_x x - i\omega t) + c.c. \tag{2.17}$$

and

$$\mathbf{e}_i = (\cos\theta\cos\varphi, \sin\varphi - \sin\theta\cos\varphi), \tag{2.18}$$

Where φ is the angle between the plane of incidence and the incident wave polarization vector. The "linear" transmitted and reflected waves are described by the usual Fresnel formulae.

The wave transmitted into the crystal gives rise to nonlinear polarization, according to the expansion (2.5)–(2.7). The polarization dependence of the reflected SH(TH) signal will be determined by the fact that the amplitude of harmonic wave appears to be proportional to the projection of nonlinear source vector on the direction of the polarization of the harmonic being registered. The final form the reflected harmonic field E_H (with frequency ω_H) is as follows [19, 220]:

$$E_{H,s} = (E_H)_y = i\frac{4\pi\omega_H^2}{c^2(k_{1z}+k_{2z})}\,P_y^{NL}(\omega_H);$$

$$E_{H,p} = i\frac{4\pi\omega_H^2 k_1}{c^2(k_2^2 k_{1z}+k_1^2 k_{2z})}(k_{2z}P_z^{NL}+k_{2z}P_x^{NL}); \tag{2.19}$$

here the subscripts s, p on the amplitudes of harmonic fields refer to the case of s ($\varphi = \pi/2$) and p($\varphi = 0$) polarization of these fields, indices 1,2 on the wave vectors of the harmonics correspond the vacuum and nonlinear medium

$$k_1 = \frac{\omega_H}{c} \,, \quad k_2^2 = \left(\frac{\omega_H}{c}\right)^2 \epsilon(\omega_H) \,.$$

In the simplest case of the second harmonic generation in reflection from a non-centrosymmetric crystal the main contribution to the nonlinear polarization P^{NL} (2ω) on the right hand side of Eq. (2.19) is made by the bulk dipole quadratic source given by the first term in expansion (2.6) (see [105, 118, 122]). In this case information on the crystalline surface structure is contained in the dependence of SH intensity upon the direction of the polarization of the incident wave with respect to the plane of incidence, $I_{SH}(\varphi)$, and upon the orientation of the crystal axis relative to the same plane, $I_{SH}(\psi)$.

For instance, SHG in reflection from the plane (100) of the crystals of class $\bar{4}3m$ (GaAs, GaP) in which only one component of quadratic dipole susceptibility differs from zero (see (2.11)) is characterized by the following dependence on φ and ψ — the angles between the plane of incidence and the [010] axis:

$$I_{SH}(\varphi, \psi) \sim (P^{(2)D}, e_s)^2 \sim |\chi_{123}^{(2)D}|^2 |\cos\theta \, \cos 2\psi \, (1 + \cos 2\varphi) -$$
$$- \sin 2\varphi \, \sin 2\psi \,|^2 \,, \tag{2.20}$$

$$H_{SH}(\varphi, \psi) \sim (P^{(2)D}, e_p) \sim |\chi_{123}^{(2)D}|^2 |\cos\theta \, \sin 2\psi +$$
$$+ \cos 2\varphi \, (\cos\theta \, \sin 2\psi - \cos 2\psi)|^2 \,; \tag{2.21}$$

here e_p, e_s are unit vectors referring to p-polarization (in the plane of incidence and s-polarization perpendicular to it), θ is the angle of the refraction.

Under the condition that the wave of the fundamental frequency is strictly s- or p-polarized, the dependence $I_{SH,s,p}$ reflects the symmetry of the (100) plane.

The second harmonic generation in reflection from centrosymmetric crystal or isotropic medium is much less efficient process, since it is determined by either relatively weak bulk quadrupole susceptibilities or dipole susceptibilities of several surface atomic layers.

We note first of all that, as follows from (2.13), (2.10), the second harmonic signal in reflection from isotropic medium is polarized almost entirely in the plane of incidence both in the cases of s- and p-polarized wave at the fundamental frequency (see below).

Thus, the appearance of s-polarized component in the second harmonic wave reflected from the centrosymmetric crystal at $\varphi = 0$ or $\varphi = \frac{\pi}{2}$ must be attributed mainly to the anisotropy of its quadrupole susceptibility: $E_{SH} \sim \zeta_Q$.

However, a more detailed analysis [221, 222] shows that, in general, there is a nonvanishing s-polarized component of the nonlinear polarization even at the reflection from an isotropic medium; the effect is due to existence of the surface nonlinearity. For the case of the reflected surface SHG we have:

$$E_s(2\omega) = i\, 4\pi \left(\frac{2\omega}{c}\right)^2 \frac{1}{k_{1z}+k_{2z}} \, 2\chi_{yzy}^{(2)S} \cdot$$

$$\cdot \left[1 - \frac{\sin(\theta_1 - \theta_2)}{\sin(\theta_1 + \theta_2)}\right] E_s(\omega)\, E_p(\omega)$$

where θ_1, θ_2 are the angles of incidence and refraction of the fundamental beam, and $E_s(\omega)$, $E_p(\omega)$ are the s- and p-components of the input field in medium 1 (vacuum).

However, the so-called rule of "s- s ban" [225] (i.e. the absense of s-polarized SH light with s-polarized fundamental beam) is strongly obeyed in the case of a centrosymmetric isotropic medium with a smooth boundary. The experimentally observed violation of the rule is tentatively due to the roughness of real surfaces [225].

Anisotropy of the bulk quadrupole quadratic (as well as dipole cubic) susceptibility in crystals has another important manifestation. The nonvanishing of the parameters $\zeta_Q(\zeta_D)$ gives rise to the appearance of the SH intensity dependence on the crystal orientation with respect to the plane of incidence, i.e. the dependence of the form $I_{SH,TH} = I_{SH,TH}(\psi)$. It is clear that there is no such dependence in reflection from an isotropic medium.

The results of calculations of the dependences of SH intensities $I_{SH}(\psi)$, $I_{TH}(\psi)$ for cubic centrosymmetrical crystals of m3m class reflected from the (100) or (111) planes in different polarization configurations are given in Table II. As one can see, the dependencies $I_{SH}(\psi)$, $I_{TH}(\psi)$ indeed reflect the symmetry of the surface from which nonlinear reflection occurs.

In discussing the probing of the state of the crystal surface with the help of nonlinear optical reflection, we have not so far discussed the question of surface layer depth, contributing to the reflected harmonic, i.e. the depth of probe penetration into the substance under investigation.

To give a quantitative answer to this question it is necessary to solve the problem of harmonic generation in reflection from a thin crystal layer taking into account the transmission of the incident wave into the layer and reflection from the back face, then let d- the thickness of this layerd approach infinity and determine the values of $d > d_0$ for which the result is independent of the thickness. The answer is that upon reflection from a transparent or weakly absorbing medium in which case the bulk nonlinear (either dipole or quadrupole)

Table II.

Dependence of the intensity of the reflected second (SH) and third (TH) harmonics on the angle ψ of rotation of the crystal around the normal to the surface (cubic crystal of class m3m).

Surface of the crystal from which the reflection occurs	Polarization of incident wave	Polarization of the harmonic	The dependence $I_{SH} = I_{SH}(\psi)$	The dependence $I_{TH} = I_{TH}(\psi)$
(100)	p	s	$|\zeta_Q|^2 \sin^2 4\psi$	$|\zeta_D|^2 \sin^2 4\psi$
	s	s	$|\zeta_Q|^2 \sin^2 4\psi$	$|\zeta_D \cos 4\psi + 4\chi_{1111}^{(3)D} - \zeta_D|^2$
	s	p	$|\zeta_Q \cos 4\psi + 8\chi - \zeta_Q + a\chi^s|^2$	$|\zeta_D|^2 \sin^2 4\psi$
	p	p	$|\zeta_Q \cos 4\psi + b\chi + c\zeta_Q + d\chi^s|^2$	$|\zeta_D \cos 4\psi + q\zeta_D + e\chi_{1111}^{(3)D}|^2$
(111)	p	s	$|\zeta_Q|^2 \sin^2 3\psi$	$|\zeta_D|^2 \sin^2 3\psi$
	s	s	$|\zeta_Q|^2 \sin^2 3\psi$	$\left|\chi_{1111}^{(3)D} - \dfrac{1}{2}\zeta_D\right|^2$
	s	p	$|\zeta_Q \cos 3\psi + f\zeta_Q + g\chi + h\chi^s|^2$	$|\zeta_D|^2 \sin^2 3\psi$
	p	p	$|\zeta_Q \cos 3\psi + k\zeta_Q + l\chi + m\chi^s|^2$	$|\zeta_D \cos 3\psi + n\zeta_D + r\chi_{1111}^{(3)D}|^2$

1) The coefficients a, b, c, . . . , are functions of the angle of refraction θ. 2) The general factors depending solely on the angle θ are not written out in the formulas. 3) χ^s denotes the linear combination of the susceptibilities $\chi_{\parallel\parallel\parallel}^{Ds}$, $\chi_{\perp\perp\perp}^{Ds}$, $\chi_{\parallel\perp\parallel}^{Ds}$, $\chi \equiv \chi_{1212}^{(2)Q}$.

polarization plays the dominant role, $d_0 \approx \lambda/4\pi n$ where λ is the probing wavelength [122]. If the reflection occurs from strongly absorbing medium, then

$$d_0 \approx \gamma_\omega^{-1} + \gamma_H^{-1}, \tag{2.22}$$

where $\gamma_\omega, \gamma_H (\geqslant \lambda^{-1})$ are the optical absorption coefficients on the fundamental and harmonic frequencies respectively. This result can be easily interpreted since the reflected harmonic is generated by the radiation of nonlinear oscillators of the second medium backwards into the linear medium. If the absorption occurs at the fundamental or harmonic frequencies, the oscillators lying at depths exceeding d_0 (2.22) cannot contribute to the reflected signal.

The above considerations refer to the bulk contribution to the reflected harmonic since the depth of the layer providing surface nonlinear dipole response, as it has already been mentioned, is equal only to several lattice constants.

Although so far we have discussed, for the concreteness, harmonic generation in reflection, the main results of the analysis may be also used in the case of the generation of sum frequencies (SFG) or difference frequencies (DFG) and combination optical frequencies, when incident light field contains two or more frequency components. The main difference in the latter case from SHG and THG consists in the possibility to fixing different polarization states of different spectral components of the pump field, as well as in the possibility of selecting more disirable frequency range in which the surface is probed and the signal is recorded.

The observation of the bulk dipole second harmonic generation in reflection from centrosymmetric metals and semiconductors in the presence of the static electric field [124, 125, 223] is the particular case of sum (or combination) frequency generation. Actually in these experiments one observes the process of mixing of three frequencies of the type $2\omega = \omega + \omega + 0$, where "0" corresponds to the static field "frequency". This process at the surface of controsymmetric media is described by dipole cubic susceptibilities $\chi_{ijkl}^{(3)D} (2\omega,\omega,\omega,0)$ (compare (2.7)).

2.2.4. Surface and bulk nonlinearities contributing into the reflected harmonic generation

As it was mentioned above in the surface SHG experiments, the detector unavoidably collects the total second harmonic signal contributed by both the surface and the bulk. In fact, as far as the signal is concerned one can define an equivalent surface nonlinear polarization as a source for SHG to take into account contributions from both the surface and the bulk [19, 221, 224].

However, difficulty may arise if one would like to separately determine the two contributions from measurements. Only in special cases with proper combinations of beam polarizations will this possible.

One simple polarization configuration giving an opportunity to strictly isolate the surface nonlinearity contribution into reflected SHG has been already discussed above for the case of the reflection off an isotropic medium: the s-polarized component of the reflected SH field is entirely induced by the surface nonlinearity.

Another important particular case which gives an opportunity to distinguish the bulk and the surface contributions into nonlinear source is as follows [221]. If one uses two orthogonally polarized fundamental input beams with slightly different frequencies propagating collinearly in the centrosymmetric isotropic nonlinear medium, the sum frequency bulk nonlinear polarization will vanish, as it follows from the eq. (2.13). Then only the surface nonlinearity should contribute to the sum frequency signal. By choosing different polarization contributions while keeping the polarizations of two fundamental beams orthogonal, all the nonvanishing surface susceptibility elements $\chi^{(2)S}_{ijk}$ can be measured.

In the case of a crystalline medium, both the surface and the bulk nonlinearities could acquire some anisotropic terms but separate determination of the surface and bulk anisotropic terms is possible from measurements of SHG with a few different crystalline surfaces. This has been demonstrated in Si and Ge [19, 225].

With a somewhat more complicated arrangement, it is also possible to eliminate the bulk anisotropic contributions while measuring the surface anisotropy. For example, for a Si (111) – 1 x 1 or 7 x 7 reconstructed surface, both the surface and the bulk contributions have a 3m symmetry. The surface and the bulk anisotropic terms in the nonlinear polarization are, respectively:

$$P^{(2)S} = \chi^{(2)S}_{xxx} \Sigma e_\xi E_\xi(\omega) E_\xi(\omega)$$

and

$$P^{(2)V}_A = \eta \Sigma e_a E_a \nabla_a E_a$$

where e_ξ are along the directions of the projected [111] axes on the surface, and e_a are along the principle axes of the zine-blende crystal. To make $P^{(2)V}_A$ vanish, one can employ two counter propagating fundamental beams in Si. A Breuster angle of incidence with p-polarization can be used to eliminate the interference effect due to reflection of the fundamental beam into the Si bulk. The three-fold symmetry thus observed in SHG should arise entirely from the surface anisotropy.

One more possibility to distinguish the surface and the bulk nonlinearities is connected with resonant enhancement of the former. For example, due to the reconstruction of the Si (111) − 7 x 7 or 1 x 1 surfaces, a new Si surface absorption band in the near infrared occurs in the wavelength range $1 - 2\ \mu m$, associated with the dangling Si bond formation of the reconstructed crystal structure, thus leading to resonant enhancement of the surface contribution to the SHG signal with the incident beam from Nd : YAG laser ($\lambda = 1.06\ \mu m$) [219, 226] (see also [261, 262]).

While it is easy to see that the strong surface nonlinearity arises from the broken symmetry at the interface, the more detailed physical origin of the surface nonlinearity is still a subject of interest. Is the field discontinuity [116] or the structural discontinuity [19] or both at the interface responsible for the nonlinearity? Is the nonlinearity dominated by local or nonlocal response of the interfacial layer to the applied field?

In an important earlier work by Bloembergen et al. [116], the electroquadrupole (nonlocal) contribution to the surface nonlinearity resulting from the rapid variation of the normal component of the electric field across the interface was stressed. This led to the belief that surface SHG would be insensitive to molecular adsorbates on metal surfaces, and only weakly sensitive to adsorbates on low refractive index materials. On the other hand, recent experiments on better characterized surfaces have shown that SHG can be a sensitive probe of the surface structure and molecular adsorbates on surfaces [227]. They indicate that the dipole (local) contribution to the surface nonlinearity in many cases may actually be important [19]. Which surface nonlinear contribution either dipole (local) or electric quadrupole or magnetic dipole (nonlocal) − may dominate would depend on the characteristics of the particular surface system in question.

In the recent paper by Shen et al. [222] the results of the first experiments in sorting out the local and nonlocal contributions to the surface nonlinearity for surface SHG are discussed. According to theory, the local contribution depends only weakly on the dielectric constants of the bounding media at the interface. The nonlocal contribution, on the other hand, is highly sensitive to the difference of the two dielectric constants, becoming vanishingly small when the two are matched.

Thus, experimentally, the two contributions can be most easily sorted out by varying the dielectric constants of one of the media. The local contribution can be strongly enhanced when SHG is at resonance with an allowed transition involving a surface state, such as the excited state of an adsorbed molecule or the surface electronic states involving dangling bonds on a clean reconstructed surface.

Electric dipole contribution from the first monolayer of adsorbed molecules at surfaces of centrosymmetric materials to the reflected SH light is now widely used to characterize the process of adsorption [109, 110, 226, 228–239]. The effect was demonstrated for several systems, such as rodamine 6G at the fused silica/air interface [229, 236], including monolayer detection of these molecules with the help of SHG of a 20 mW cw diode laser [231], p-nitrobenzoic acid at the interface with air or ethanol [230], silver electrodes covered with pyridine and pyridazine [109], Langmuir-Blodgett films on the liquid surface [237, 238], etc. It was shown that the spectral shape of the resonance could identify the adsorbed molecules and that polarized measurements could yield information about molecular orientation at the surface [226]. Unlike other surface-sensitive techniques using electron scattering (LEED) or photoelectron spectroscopy (UPS, XPS) for this purposes the nonlinear optical method is applicable to dense media, not restricted to ultra-high vacuum conditions, and it provides relatively high spectral, spatial and time resolution. Also, there exists a possibility of the study of rough metal surfaces by using surface-enhanced SHG [240–242, 263, 267] a nonlinear coherent analog of SERS.

2.3. Study of the states and fast phase transformations of semiconductor surfaces by means of the generation of optical harmonics and sum frequencies on reflection

2.3.1. Experiments on second harmonic generation in reflection for the study of the pulsed laser annealing dynamics in GaAs

A block diagram of the experimental arrangement [18] for studying of PLA dynamics of GaAs surface in shown in Fig. 6. The annealing was accomplished by ruby laser pulses ($\tau_p \approx 30$ ns); the uniformity of the intensity distribution across the laser beam was obtained by a special forming system consisting of a frosted plate, a lens system and diaphragms.

The probing laser was Nd : YAG laser with simaltaneous acoustooptical Q-switching and mode locking of the longitudinal modes. The laser generated a train of 40 pulses with duration of 0.3 ns each and with a repitition period of 9.5 ns. The beam was further amplified (approximately tenfold) in a single-pass amplifier. The total energy of one "packet" of subnanosecond pulses (a typical train oscillogram is given in Fig. 7d) did not exceed 1 mJ.

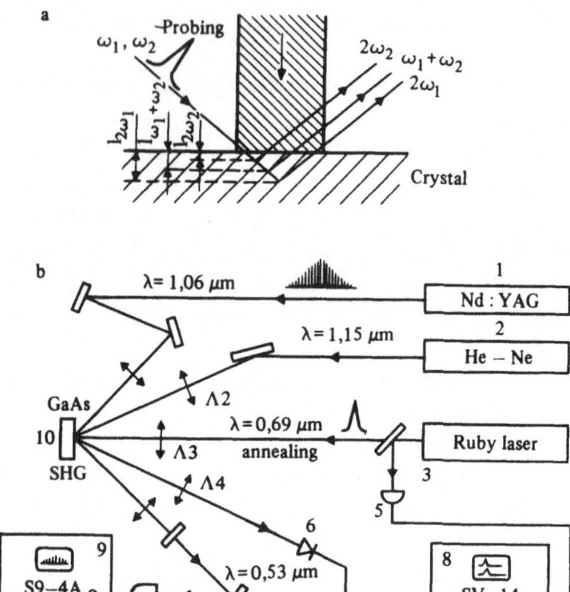

Fig. 6 a) General scheme for recording the nonlinear optical response of a semi-conductor surface subjected to pulsed laser action. The probing was performed by weak optical pulses (which do not affect the excitation process) with frequencies ω_1 and ω_2 (it is assumed that $\omega_1 > \omega_2$, in the simplest degenerate case $\omega_1 = \omega_2$); the signal of the second optical harmonic ($2\omega_1$, $2\omega_2$) and the sum frequency ($\omega_1 + \omega_2$) carry information on surface layers of different thickness: $1_{2\omega_1} = \gamma^{-1}(2\omega_1) \leqslant 1_{\omega_1+\omega_2} = \gamma^{-1}(\omega_1 + \omega_2) \leqslant 1_{2\omega_2} = \gamma^{-1}(2\omega_2)$. b) Block diagram of the experiment studying the dynamics of pulsed laser annealing of a gallium arsenide surface by recording in real-time the reflected second harmonic [18]. 1) Nd : YAG laser with active mode-locking and acoustooptical Q-switching; 2) He—Ne laser; 3) ruby laser with electrooptical Q-switching; 4) set of SZS-21 filters; 5) coaxial FK-20 photocell; 6) LFD-2A avalanche photodiode; 7) 18 ELU-FM fast photomultiplier; 8) SV-14 oscillograph (200 MHx bandwidth); 9) S9-4A oscillograph (350 MHz bandwidth); 10) GaAs sample.

The annealing pulse and a train of probing pulses were synchronized in time by electric synchronization of the corresponding Q-switches.

The radiation of continuous He—Ne laser with wavelength $\lambda = 1.15$ μm was used for simaltaneous recording of high reflection phase (HRP).

SH signal due to probing beam reflection from the sample surface was recorded with the help of a fast photomutiplier (18ELU-FM) and C9-4A oscillograph with transmission band 350 MHz. A two-beam recording oscillograph C8-18

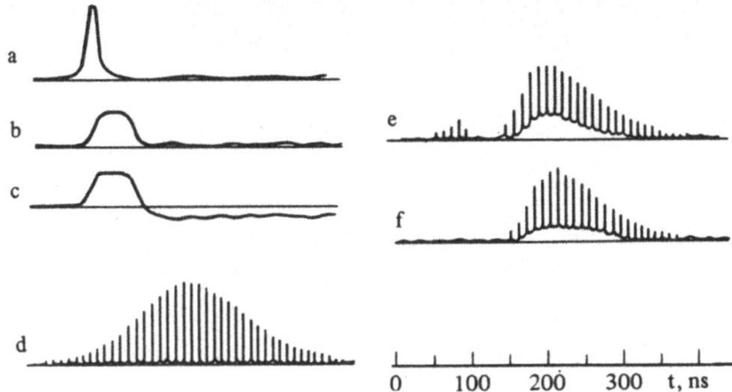

Fig. 7. Oscillograms of the annealing ruby-laser pulse (a), of the reflected probe
pulses (b, c), of the probing pulse (d), and of the SH pulses (e, f) from
the unimplanted (b, e) and implanted (c, f) sections of the (110) surface
of the GaAs crystal. The sweeps in all oscillograms are synchronized.

was used to obtain time scan of HRP as well as to determine the relative position
of annealing and probing pulses in time. Gallium arsenide $(1\bar{1}0)$ — cut samples
were used. The annealing beam was incident normal to the surface, whereas
the probing pulse formed an angle 45° with the normal.

The results of the experiments are illustrated in Fig. 7, which shows the oscillograms of annealing (a) and probing (d) pulses, as well as the high reflection
phase (HRP) (b, c) and SH signal for the case of annealing of amorphous and
single-crystalline surfaces (b, f respectively).

As it is seen, SH from crystalline surface disappears during the time interval
equal to the duration of the leading edge of HRP and is absent during HRP-
plateau. This behavior of SH is easily explained on the basis of the thermal
PLA model, if we assume that the HRP-plateau corresponds to the liquid melted
state of the semiconductor. Estimates show that SH intensity should not drop
so drastically in the plasma PLA model.

In fact, let HRP correspond to dense electron-hole plasma in a solid state. Then
the change of SH signal will be affected by the following two factors: 1) decrease
in the density of bounded electrons n_0, which determine mainly the value of
$\chi^{(2)}$; 2) additional absorption of probing radiation and the SH signal by free
carriers (screening by the carriers). We shall evaluate each of these factors.

The density n_0 for GaAs is equal to

$$n_0 \approx 5 \cdot 10^{23} \text{ cm}^{-3}$$

One can estimate, that the absorption of annealing pulse with typical energy
density $W = 0.4$ J/cm^2 in a layer with a thickness of $1_{abs} = 0.3 \cdot 10^{-4}$ cm leads
to the exitation of $n_e = W/1\epsilon = 4 \cdot 10^{22}$ electrons per cubic cenitimeter into
the conduction band (neglecting recombination), where $\epsilon = 3 \cdot 10^{-19}$ J is

47

the photon energy with $\lambda = 0.69$ μm. The bound electron density will be equal to $n_1 = n_0 - n_e$. Thus, relative decrease of SH will be

$$\frac{I - I_0}{I_0} = \left| \left(\frac{n_0 - n_c}{n_0} \right)^2 - 1 \right| = \left| \left(1 - \frac{4 \cdot 10^{22}}{5 \cdot 10^{23}} \right)^2 - 1 \right| \approx 0{,}16 \; .$$

Screening has a stronger effect. To take it into account we have measured the probe radiation transmission coefficient during the action of annealing pulse. With ruby laser annealing pulse the transmission at $\lambda = 1.06$ μm is measured to decrease by not more than 40%. We shall assume that this change to be entirely connected with the absorption by free carriers generated by ruby laser annealing pulse in the surface layer whose depth is determined by the absorption coefficient of this radiation; $l_{abs} = (3 \cdot 10^4 \text{ cm}^{-1})^{-1} \approx 300$ nm (Table III).

Table III.

Coefficients of optical absorption of crystalline and amorphous gallium arsenide (according to data in Refs. [131 and [132]).

λ, nm	γ_{cryst}, cm^{-1}	γ_{amorph}, cm^{-1}
1064,2	0,02	$5 \cdot 10^3$
694,2	$2,7 \cdot 10^4$	$1,5 \cdot 10^5$
532,1	$8,09 \cdot 10^4$	$3 \cdot 10^5$
354,7	$7,14 \cdot 10^5$	$7 \cdot 10^5$

Then one can give an upper limit of additional "plasma" absorption of the probing radiation: $\gamma_{1.06}^{pl} \cong 1.5 \cdot 10^4$ cm^{-1}. The change of absorption coefficient at $\lambda = 0.53$ μm owing to the appearance of plasma cannot exceed this value for $\lambda = 1.06$ μm since $\gamma_\omega^{pl} \sim [(\omega/\gamma)^2 + 1]^{-1/2}$ where $\omega = \frac{2\pi c}{\lambda}$, γ is the frequency of electronic collisions. Therefore

$$\gamma_{0,53}^{pl} < 1{,}8 \cdot 10^4 \text{ cm}^{-1}$$

In comparison with the main absorption by the bounded electrons $a_{0.53}^b = 5 \cdot 10^4$ cm^{-1} (see Table III) this increase in absorption may lead the decrease in SH intensity by not more than 50%, because

$$\frac{I_1}{I_0} \approx \left(\frac{\gamma_{0,53}^b + \gamma_{0,53}^{pl}}{\gamma_{0,53}^b} \right)^{-2} \approx 0{,}55$$

Thus, even the upper limit of the value of SH intensity decrease due to the free carriers absorption constitutes 50%, whereas in the experiment it drops by more than two orders of magnitude. This testifies to the fact that the disapperance of SH is mainly caused by the breaking of crystal lattice symmetry, i.e. melting of the lattice.

In a work [132] analytic calculations of SH intensity with taking into account the multiple reflections in the melted layer with depth d(t) (in analogy with the linear case [133, 243, 244]),using boundary conditions [105] were made to interpret quantitatively the results of these experiments on SHL at RPA on the basis of the surface melting model

$$I_{SH} = \text{Const } |R_{SH}(t)|^2 \qquad (2.23)$$

here

$$R_{SH} = \frac{\exp(i\Delta(2\omega, t) + 2i\Delta(\omega, t))}{[1 - r_{12}(2\omega)r_{23}(2\omega)e^{2i\Delta(2\omega,t)}]\,[1 - (r_{23}(\omega)r_{21}(\omega)e^{2i\Delta(\omega)})^2]} \qquad (2.24)$$

$$\Delta(x, t) = \frac{2\pi d(t)}{\lambda}\,(\epsilon_2(x))^{1/2}, \qquad \epsilon_2(\omega) = \epsilon'_2 + i\epsilon''_2$$

is the dielectric constant of melt, $\lambda = 2\pi c/\omega$,

$$r_{\alpha\beta}(\omega) = \frac{(\epsilon_\alpha(\omega))^{1/2} - (\epsilon_\beta(\omega))^{1/2}}{(\epsilon_\alpha(\omega))^{1/2} + (\epsilon_\beta(\omega))^{1/2}},$$

the indices α, β refer to the vacuum (1), the melt (2) and the crystal (3).

To calculate $I_{SH}(t)$ with the help of formula (2.23) it is necessary to know the dynamics of melting, i.e. dependence d(t). This dependence for GaAs is calculated in [132] on the basis of numerical solution of nonlinear hear-conduction equation with coefficients explicitly dependent on temperature (similarly to [40, 41]). The change in the linear reflection coefficient of the surface at pumping frequency in the PLA process was also taken into account (see the caption to Fig. 8c).

The results of these calculations are presented in Fig. 8. It is evident that under the action of laser pulse with energy $W = 0.4$ J/cm^2 the melt front moves rapidly into bulk of the medium, reaches the maximum depth of the order of $6 \cdot 10^{-5}$ cm

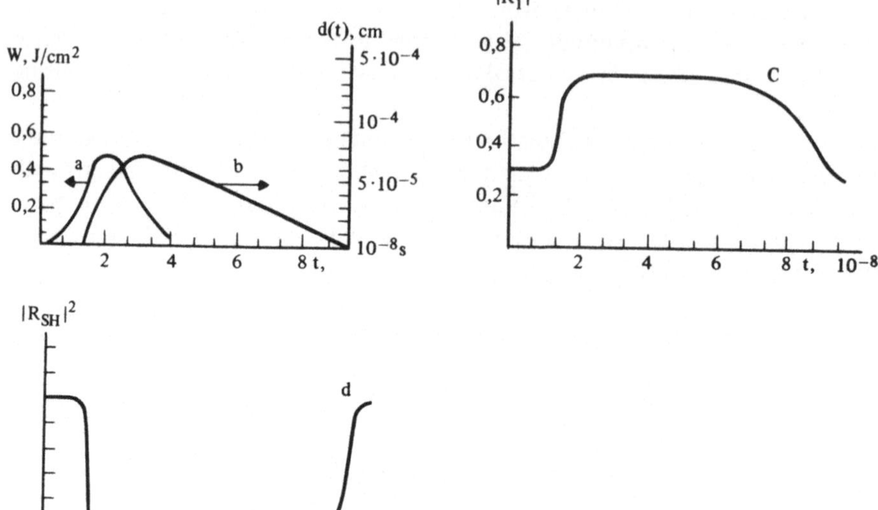

Fig. 8. Dynamics of impulsive laser annealing of GaAs (numerical calculation). a) Laser pulse with $W = 0.4$ J/cm^2, $\tau_p = 20$ ns, $\gamma = 10^{-4}$ cm^{-1}. b) The thickness of the melt d as a function of time; c) coefficient of linear reflection as a function of time ($R_1(t) = \left[r_{12}(\omega) + r_{23}(\omega)e^{2i\Delta(\omega,t)}\right]$ x x $\left[1 + r_{12}(\omega)r_{23}(\omega)e^{2i\Delta(\omega,t)}\right]^{-1}$); d) SHG as a function of time.

and returns to the surface in the process of recristallization. In the time, during which the front of the melt traverses the distance equal to the skin depth δ_{sk} (the time $\tau_{grow.}$), linear reflection coefficient increases almost twofold and then saturates to plateau (high reflectivity phase — HRP). HRP duration is determined by the roundtrip transit total time of melt front into the bulk of the crystal and back. The duration of HRP decay τ_{dec} is determined by the time of melt front transversing the skin depth δ_{sk}. Taking $\delta_{sk} = 20$ nm and estimating on the basis of Fig. 8, $\tau_{grow.} \sim 5 \cdot 10^{-9}$ sec we obtain for the melt front velocity $V_{melt} \sim 4 \cdot 10^2$ cm/sec. Estimating again with the help of Fig. 8c $\tau_{melt} \sim 10^{-8}$ sec, we get for the velocity of recristallization $V_{recris} \sim 100$ cm/sec.

The efficiency of SH generation drops at the same time that the reflection coefficient increases during the motion of the melt front into the bulk of the crystal. During subsequent recristallyation the restoration of initial SHG efficiency takes place. The characteristic times of SH decay and growth roughly coincide with τ_{gr} and τ_{dec}. This is determined by the fact that in equation (2.24) $2\Delta(\omega) \gg \Delta(2\omega)$, therefore, the dynamics of both SH as well as of the linear reflection coefficient is determined by attenuation factor of the wave with frequency $\omega(e^{-2i\Delta(\omega,t)})$. One can see, that all the linear and nonlinear optical characteristics of PLA dynamics in GaAs, given by theory and experiment, are in a good agreement with one another within the accuracy limits of the corresponding data (see also Table IV).

Table IV.
Comparison of the theoretical results of Ref. [132] with experiments [18,113] on the study of the dynamics of impulsive laser annealing in GaAs with the help of SHG in reflection.

Characteristics of ILA	Experiment	Theory
Time of growth of the phase of high reflection τ_{PHR}^{grow}, ns	< 10	5
Decay time of PHR τ_{PHR}^{dec}, ns	~ 15	14
Duration of PHR τ_{PHR}, ns	40	45
Duration of the decay of the intensity of the SH τ_{SHG}^{dec}, ns	10	5
Duration of the growth in the intensity of the SH $_{SHG}^{grow}$, ns	15	14

N.Bloembergen et al. [127] made use of the decribed above technique of SHG measurement on reflection from GaAs crystal to estimate the melting time of a crystal surface on a picosecond time scale.

In experiments [127] a single pulse of second harmonic frequency of Nd:YAG laser with duration τ_p = 12.5 ps (λ = 532 nm) was used as a probing pulse. Nonlinear reflection of this pulse incident at the angle θ = 4–5° and polarized along the $[1\bar{1}1]$ direction occurred from $(1\bar{1}0)$ face of GaAs. The SH pulse energy dependence on the density of pulse energy W_1 at the fundamental frequency (λ_1 = 532 nm) was studied. Up to values W_1 = W_{th} = 30 mJ/cm^2 (at which, according to [41], melting of GaAs surface begins) the ordinary quadratic dependence $W_{SH} = \eta W_1^2$ was observed, which is the consequence of the same dependence for SH intensity in reflection from a crystal:

$$I_{SH} = \eta_{SH}\, I_1^2 \tag{2.25}$$

following, for example, from Eq. (2.19). The coefficient η_{SH} was measured to be η_{SH} = 2.5·10^{-18} cm^2/W. However, when W_{th} is exceeded, a considerable deviation from the quadratic dependence and transition into the "saturation" regime is observed, when W increases up to W = 30 W_{th}, which is evidently connected with surface melting under the action of the leading edge of the front. The experimental dependences agree satisfactorily with theory, if in formula (2.25) we assume that

$$\eta_{SH} = \eta_{SH}(r,t) = \begin{cases} \eta_0 , & \text{for } t \leqslant t_m(r) , \\[2ex] \eta_0 \exp\left(-\dfrac{t - t_m(r)}{\tau_c}\right) , & \text{for } t > t_m(r) , \end{cases}$$

where for each point r within the irradiated spot the time moment t_m is determined from the equation

$$\int_{-\infty}^{t_m} I_1(r,t)\,dt = W_{th}$$

and time constant is chosen to be $\tau_s < 2$ psec. In this way the upper bound for the duration of GaAs surface transition to the centrosymmetrical phase is determined.

2.3.2. Diagnostics of quality of laser annealing on basis of second harmonic generation in reflection

The quality of PLA and, in particular, the degree of crystallinity of an annealed section of a previously rendered amorphous GaAs surface, may be easily estimated by measuring SHG efficiency and polarization dependence of this effect.

Indeed, SH signal intensity in reflection from single-cristalline surface of a certain orientation is the function of angles φ, θ and ψ determining the orientation of cristollographic axes of the sample, the position of the plane of incidence and probing radiation reflection plane. For instance, the typical dependencies I_{SH} on φ (which is the angle between the probing radiation polarization vector and the plane of incidence) in the case of the $(1\bar{1}0)$-cut GaS single-crystal (the angles θ and ψ are fixed), as seen in Fig. 1 from [114], exhibit extrema. These extrema I_{SH}^{max}, I_{SH}^{min} will be to a large extent smeared, if SH signal is generated from polycristalline surface, formed for example by a number of small monocristalline sections characterized by different orientation angles ψ with respect to the plane of incidence of the probing radiation.

In addition, SH intensity at the maximum of polarization dependence for the annealed region, scaled to the corresponding SH intensity from unimplanted (and, hence, unannealed) crystalline section

$$\eta_{SH}^{PLA} = I_{SH}^{PLA} / I_{SH}^{cryst} \tag{2.26}$$

can also characterize the degree of crystallinity restoration of the surface layer as a result of PLA.

Experiments on determining PLA quality on the basis of SHG were carried out on the experimental arrangement shown in Fig. 9.

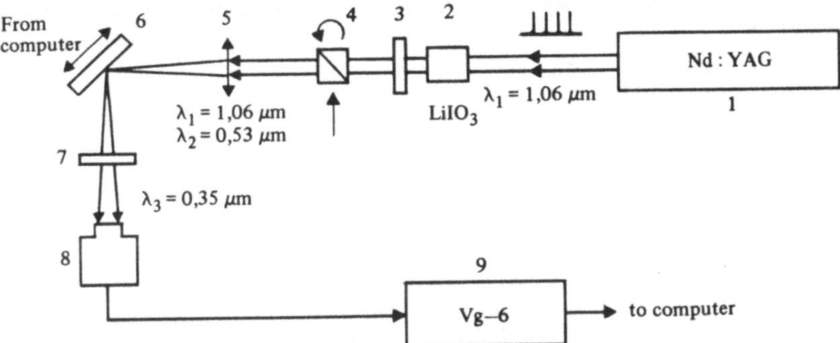

Fig. 9. Block diagram of the experiment for studying with the help of SHG and SFG the degree of amorphousness of a GaAs surface produced by ion implantation and the quality of its laser annealing. 1) Probing Nd:YAG laser with active modelocking and acoustooptical q-switching; 2) LiIO$_3$ crystal for frequency doubling (used only in experiments on SFG); 3) ZhS-4 filter (cuts off the parasitic radiation with $\lambda_{SF}=351.7$ nm); 4) polarization rotator; 5) lens; 6) sample; 7) set of SZS-21 light filters (for SHG) or UFS-2 light filters (for SFG); 8) photomultiplier FEU-106; 9) selective amplifier.

The beam of quasicontinious garnet raser with neodymium operating with a repetition frequency 1 kHz was used as a probe radiation. The recording system consisted of a FEU-106 radiation detector, box-car integrator V9-5 and an automated complex for collecting and processing data on the basis of microcomputer "Electronica-80". It enabled to detect weak signals with great signal-to-noise ratio.

The mechanically polished ($1\bar{1}0$)-cut GaAs samples implanted by phosphor P$^+$ ions with a dose D = $6 \cdot 10^{15}$ cm^{-2} and energy E = 40keV annealed by a series of ruby laser pulses (τ_p = 30 nsec) with different energy densities: W = 0.27, W = 0.21, W = 0.16, W = 0.08, W = 0.05 J/cm^2 were investigated.

It should be noted that S H signal from amorphous layer was approximately 80 times less intense than from nonimplanted section of a single crystal. The incomplete disappearance of SH may be linked to incomplete disordering of the surface layer of the implanted area. Nondisordered deep crystal layers could also contribute to the residual SHG signal since the "probe" penetration depth was of the same order of magnitude as the depth of amorphous layer, the latter value, according to [31] is equal to $\sim$ 50 nm, while the probing

53

depth, limited by SH radiation absorption in GaAs amorphous layer is equal to $l_{probe} \approx \gamma_{2\omega}^{-1} \approx 30$ nm, according to Table III. The thickness of the layer in which the annealing radiation is absorbed is equal to ~ 60 nm.

The results of the experiment (dependence $I_{SH}(\varphi)$) are given in Fig. 10, from which it is seen that SH does not indicate-polycrystallinity for all energy densities of the annealing radiation. At the same time a strong dependence of signal amplitude on annealing pulse energy is observed.

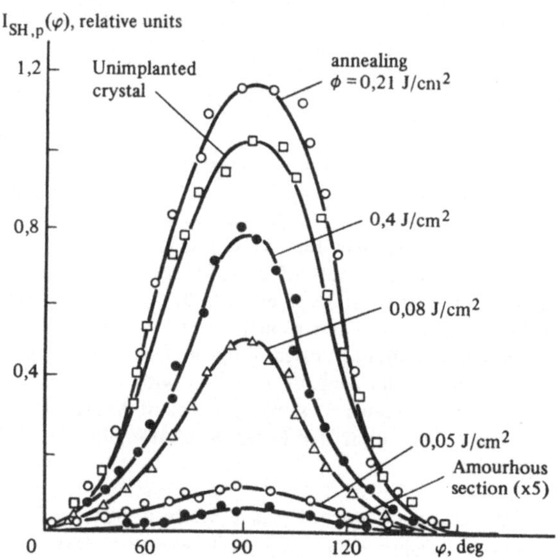

Fig. 10. Polarization dependence of SH on the section of the (110) surface of GaAs subjected to PLA with a different dosage of the annealing radiation.

The control of the PLA quality of centrosymmetric crystals such as Si or Ge demands a slightly different attitude (see below, Sec. 2.3.3).

2.3.3. Determination of degree of disordering of a crystal surface from third harmonic and second harmonic generation in reflection

We have already called attention to the high SHG and SFG sensitivity to the presence or absence of amorphous surface layer in the noncentrosymmetrial GaAs crystal. It is clear that these effects, as well as THG (see below) can be also used for quantitative estimation of the degree of partial disordering of this layer. Moreover, as will be shown below, nonlinear optical methods are

54

much more sensitive and accurate, more highly localized, and are simpler than other methods for analyzing the degree of amorphousness of the surface used thus far — Rutherford backscattering of light ions, diffraction of slow electrons, X-ray diffraction analysis, etc.

It can be easily seen, however, that characteristics of non-linear-reflected signals, which are to be used for studying surface perfection degree, must be different for noncentrosymmetrical crystals (GaAs, GaP, InSb etc.) and crystals possessing the centre of symmetry (Si, Ge etc.). In analogy to (2.26), the intensity of the second harmonic of the optical radiation, probing the disordered section of the crystal surface I_{SH}, scaled to the intensity of the SH from an undamaged section of the same crystal, I_{SH}^{cryst}, can serve as a characteristic of the degree of disordering of noncentrosymmetric crystals:

$$\eta_{SH} = I_{SH}^{a} / I_{SH}^{cryst} \qquad (2.27)$$

In our experiments on SHG we studied (100)-cut GaAs samples, whose surface was to a certain degree rendered amorphous by implantation of 80 keV Te^+ and S^+ ions with energy 80 keV with an irradiation dose ranging from $1 \cdot 10^{12}$ to $6 \cdot 10^{15}$ cm^{-2} [113]. The GaAs surface was mechanically polishend and chemically etched prior the ion bombardment. The polarization dependences of SH signal from the partially disordered surface were similar to those from the undamaged section (and in the latter case they coincided with dependencies computed from the formulas [105] see also Sec. 2.2). The intensity of SH from the implanted part of the crystal surface, however, was 1 to 2 orders of magnitude lower than from crystalline surface-depending on implantation dosage.

Fig. 11 presents dependence of η_{SH} on the dosage D, which is, as can be seen, quite distinct. It reflects the loss of long-range order of the initially noncentrosymmetrical crystal with the increase of implantation dosage.

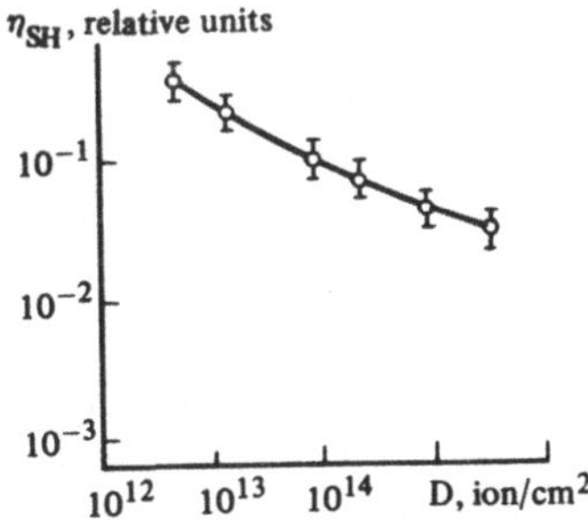

Fig. 11. The degree of amorphousness of the (100) surface according to SHG η_{SH} (λ_{SH} = 532 nm) as a function of the dosage of surface implantation by Te^+ ion with an energy of E = 80 keV.

Even stronger dependence on dosage D reveals the value of relative efficiency of sum frequency generation

$$\eta_{SF} = I^a_{SF} / I^{cryst}_{SF} \tag{2.28}$$

here I^a_{SF} is the intensity of sum frequency signal (SF) with wavelength $\lambda_{SF} =$ = 351.7 nm, obtained from GaAs section rendered partially amorphous by ion bombardment, with a help of mixing the fundamental Nd : YAG laser radiation (λ_1 = 1064.2 nm) and its second harmonic (λ_2 = 532.1 nm), generated beforhand in LiIO$_3$ crystal (the waves mixed on GaAs surface were linearly polarized in othogonal directions); I^{cryst}_{SF} is the intensity of SF signal from the undamaged region of crystal surface.

The dependencies $\eta_{sf}(D)$ for a number of GaAs samples are shown in Fig. 12.

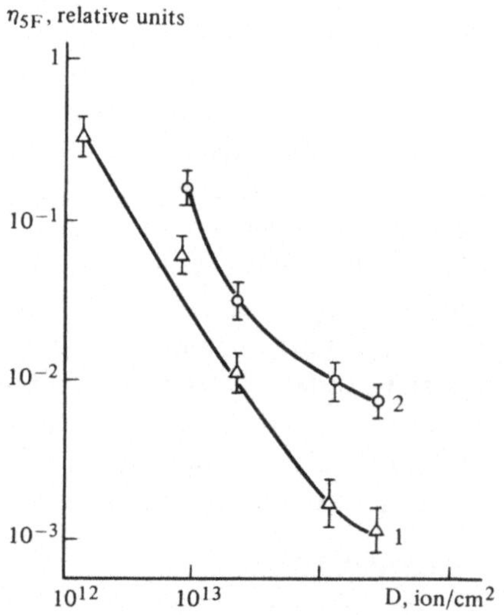

Fig. 12. The degree of amorphousness of the (100) surface of GaAs according to SFG, (λ_{SF} = 351.7 nm) as a function of the dosage of surface implantation by Te$^+$ (1) and S$^+$ (2) ions with an energy of E = 80 keV.

The difference between $\eta_{SH}(D)$ and $\eta_{SF}(D)$ is linked primarily to the difference between the probing depths: in case of SH $l_{SH} \approx \gamma^{-1}_{SH} \approx 50$ nm, in case of SF $l_{SF} \approx \gamma^{-1}_{SF} \approx 15$ nm. The latter circumstance is of prime importance, since it allows to obtain data on the degree of disordering of very thin surface layers. The quantaties η_{SH}, η_{SF} can be easily and accurately measured; they characterize quantitatively the degree of amorphousness of the surface layer of a non-symmetric crystal.

56

By means of SFG one can easily record GaAs surface amorphization caused by such low implantation dosages ($< 10^{12}$ cm^{-2}) that cannot be detected by other methods used thus far.

In particular, Fig. 13 shows the dependence of η_{RBS} — a parameter determined from the comparison of Rutherford back scattering spectra (RBS) of helium nuclei from implanted and unimplanted sections of crystal surfaces on the implantation dosage [31]. The measurements were performed on the same samples as in the case of SHG and SFG measurements (Fig. 11–12). It follows from the comparison of Fig. 11–13 that the dependence $\eta_{RBS}(D)$ is much less distinct than $\eta_{SF}(D)$ or even $\eta_{SH}(D)$.

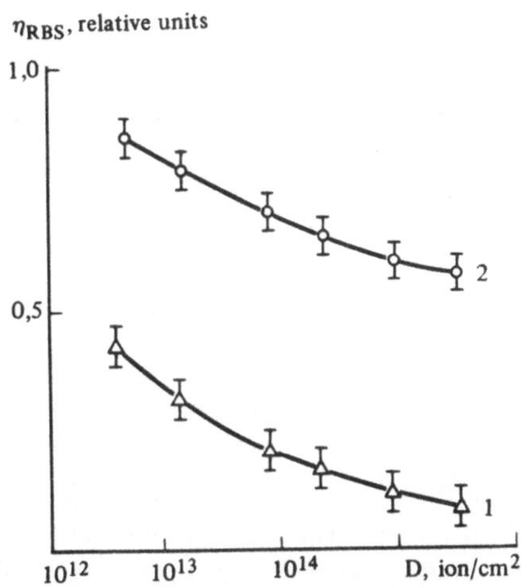

Fig. 13. The degree of amorphousness of the (100) surface of GaAs according to Rutherford backscattering η_{RBS} as a function of the dosage at implantation by Te$^+$(1) and S$^+$ (2) ions with energy E = 80 keV.

By means of SFG in reflection from the surface, subjected to "laser amorphization" one can detect "amorphization rings" produced by irradiating monocrystalline GaAs (100) surface by short laser pulses with λ = 532 and 265 nm, duration τ_p = 30 ps and fluencies W = 0.05 – 0.1 J/cm^2 (see Fig. 14) [113]. The presence of the amorphous ring is easily recorded by a sharp decrease of SF signal intensity in reflection from the corresponding area. The polarization dependence of a weak SH signal from the section of the laser induced amorphous ring appeared to coincide with that from the undamaged region. This indicates a partial preservation of the long range order in the laser induced amorphous layer which is characteristic for the starting crystal.

57

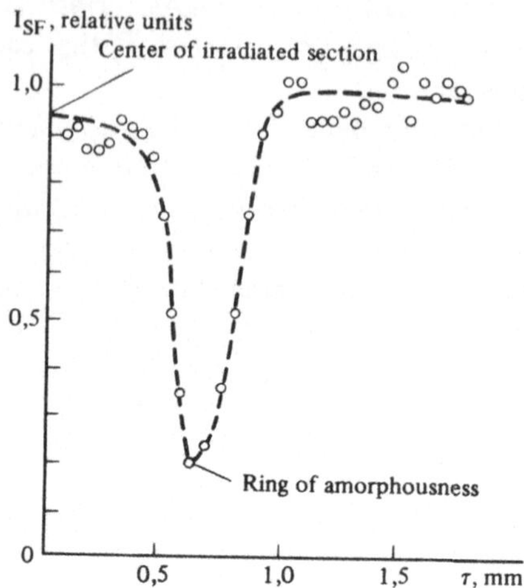

Fig. 14. The amorphous ring along the circumference of a section irradiated with picosecond laser pulse with $\lambda = 266$ nm of the (100) surface of GaAs, observed with the help of SFG in reflection.

In studying crystals with centrosymmetric lattice (further for definitness we'll speak about crystalline silicon) the relative intensity of the reflected SH can not be considered as a reliable quantitative criterion of surface perfection since not only ordered crystals but amorphous media as well exhibit quadrupole susceptibilities of the second order $\chi_{ijkl}^{(2)Q}$, describing SHG effect in these media. However, the value of anisotropic (i.e. independent of crystal orientation with respect to the plane of incidence) contribution to the intensity of SH, as well as SF and TH could serve for this purposes [247–250].

The scheme of the experiments [249, 250] is similar to the one given in Fig. 9 with Si sample with plane (111) being substituted for GaAs crystal as a nonlinear mirror. The sample was rotated about the (111) axis by a stepper motor. The surface was implantated with phosphor ions with energy 80 keV.

The SH dependence on the rotation angle of the crystal around the normal to the surface ψ is, according to Table II, the result of interference of anisotropic ς_Q and isotropic B contributions:

$$I_{SH}^{p \rightarrow p}(\psi) \sim |\varsigma_Q \cos 3\psi + B|^2 \qquad (2.29)$$

Since in isotropic medium $\varsigma_Q = 0$, it is natural to expect the anisotropic contribution to decrease with the increase of the degree of disordering of the crystalling lattice as the implantation dosage D is increased.

58

Fig. 15 presents the measured dependences $I_{SH}(\psi)$ at different values of
the dosage D. The character of this dependence for pure Si (Fig. 15a) — six
peaks in the interval $\psi = 0 - 360^o$ — indicates that the anisotropic contribution
dominates ($|\zeta_Q| > |B|$).

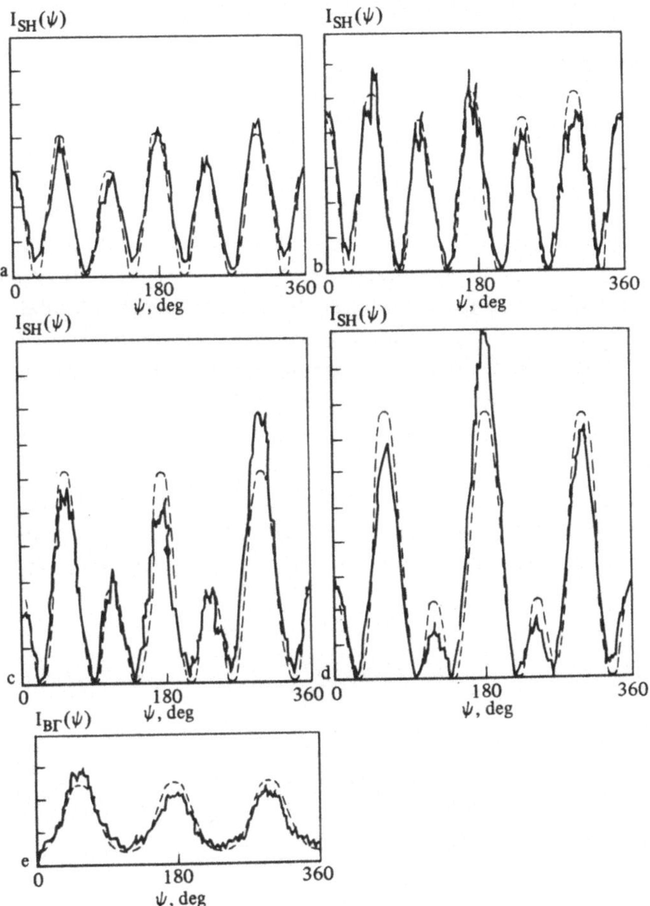

Fig. 15. The intensity of the quadrupole second harmonic ($\lambda_{SH} = 0.53\ \mu m$),
reflected from the (111) face of silicon in the p$\longrightarrow$p scheme as a function
of the angle ψ of rotation of the crystal around the [111] axis with
different dosages of surface implantation by phosphorus ions (E = 80keV):
The solid lines show the experiment and the broken lines show the curves
calculated using the formula (2.26) for five values of the ratio ξ_Q/B;
the implantation dosage (10^{13} ions/cm^2) = 0 (pure Si); $\xi_Q/B = 11$ (a),
1.8, 9,0 (b); 3.6; 5.0 (c), 12; 2.3 and 24; 0.4 (e).

59

At $D < 2 \cdot 10^{13}$ cm^{-2} no noticeable changes of the dependence I_{SH} (ψ) occurs. With further dosage increase (Fig. 15c and d) redistribution of the magnituds of the peaks is observed, indicating the increase of isotropic contribution to SH signal. Finally, for $D = 2.4 \cdot 10^{14}$ cm^{-2} the dependence I_{SH} (ψ) has three peaks, i.e. $|\varsigma_Q / B|$.

Thus, the quantity

$$\eta_{SH}^Q = |\varsigma_Q / B| \qquad (2.30)$$

measured for some fixed parameters of the experiment (at a fixed angle θ, for example) can quantitatively characterize the degree of perfection of a crystalline structure of centrosymmetrical crystals. Above introduced nonlinear-optical quantity of surface ramdomization can be compared with other values, characterizing the same process, and can be used to check a model of surface randomization suggested by Gibbons [251]. In this model each ion knocks atoms out of its sites and creates a large number of radiation defects of various types. The latter are localized in a tunnel-shaped volume along the ion trace in crystal having a cross-section σ_i However, in order to achieve complete amorphization of a material in a tunnel, energy deposition per unit volume should exceed a definite value. In the case of light ions this value is achieved when several tunnels are superimposed. With this concept in mind Gibbons suggested a formula for relative area of the amorphous material depending on the dose of ion implantation

$$S = 1 - \sum_{k=0}^{m} \frac{(\sigma_i D)^k}{k!} e^{-\sigma_i D} \qquad (2.31)$$

where (m + 1) is a number of superimposed tunnels necessary to achieve complete amorphization. This value of S can be related to the value S′ which describes a relative value of the isotropic contribution to SH signal

$$S' = 1 - \eta_{SH}^Q / \eta_{SH,0}^Q \qquad (2.32)$$

where η_{SH}^Q is determined in (2.30) for an ion-implanted crystal, and $\eta_{SH,0}^Q$ is the same value for the pure non-implanted silicon.

The dependence of S′ on the dose of ion implantation D, calculated from our experimental data [252] is shown by circles in Fig. 16. It should be compared with a computer simulation of the dependence S(D) using (2.31). The σ_i and m values were varied to achieve best fit to the experimental points. Change of σ_i causes the curve S(D) to shift along D-axis as a whole while variation of m determines the initial tilt of the curve. The values obtained from numeric fit are m + 1 = 3 and σ_i = 60A^2. Thus, we can estimate the critical energy of amorphization

$$E_a \approx (m+1)\, E/\sigma_i l$$

where l is the mean pass of an ion in a crystal. With E = 80 keV and l $\approx$ 1000A we obtain E $\approx 4 \cdot 10^{21}$ keV/cm^3 which is in good agreement with other data. Similar results were received in [134, 247, 248].

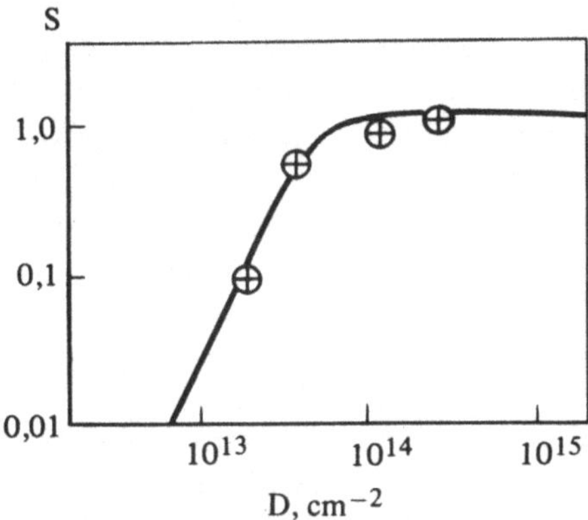

Fig. 16. Surface randomization of Si. Points-experiment, curve — model calcualtion.

It should be noted that anisotropy of SH signal in reflection from (100) and (110) faces of crystalline silicon was first discovered by Driscoll et al. [129, 130], who used single pulses from Nd : YAD mode-locked laser pules (pulse duration τ = 30 ps). They reported a high efficiency of SHG process in reflection from (111) plane in Si (only an order of magnitude lower than for reflection from a noncentrosymmetrical GaP crystal) and they revorded the dependence of S-polarized SH with s-polarized pump of the form $I_{SH}^{s-s}(\psi) \sim Sin^2 3\psi$ and of p-polarized SH at p-polarized pump of the form $I_{SH}^{p-p} \sim (1 - \cos 3\psi)^2$, where ψ is the angle of the rotation of the crystal relative to the plane of incidence of the probing beams. One must note, that in earlier experiments by Bloembergen et al. no dependence of SH signal in Ge and Si on the crystal orientation relative to the plane of incidence was observed.
Somewhat later Shen and his co-workers [19] confirmed the experimental results obtained by Driscoll et al. showing the existence of a strong anisotropic

dependence of SH on the orientation of Si crystal in reflection from (100) and (111) planes. They also correctly interpreted these results as consequences of anisotropy of volume quadrupole nonlinear polarization of the second order.

In the case, when experiments are conducted with very clean surfaces obtained in ultrahigh vacuum by means of cleavage of oriented crystals, the presence of broken valence bonds of surface atoms gives rise to the surface structure reconstruction and, hence, causes its local symmetry to differ from the symmetry of internal crystalline planes with the same indices. In particular, a clean (111) surface of silicon, as shown in [154, 219] gives a very strong anisotropic dipole contribution to the reflected SH. With temperature of the crystal being increased, the anisotropy of this contribution changes, demonstrating surface phase transition from 2 x 1 to 7 x 7 reconstruction.

As it was stressed above, cubic dipole nonlinear susceptibility $\chi_{ijkl}^{(3)D}$ generally exhibits anisotropy just as the second-order quadrupole susceptibility $\chi_{ijkl}^{(2)}$. In particular, our experiments show the anisotropy of the tensor χ_{ijkl} (ω_a, ω_1, ω_1, $-\omega_2$) for silicon is large $|\zeta_D / \chi_{1111}^{(3)D}| \approx 1$ (when two photons ω_1 with wavelength λ_1 = 1.07 μm and one photon ω_2 with wavelength λ_2 = 1.17 m are mixed). Nevertheless, the anisotropy of tensor χ_{ijkl} ($3\omega_1$, ω_1, ω, ω_1) describing the generation of dipole third harmonic of the fundamental Nd : YAG laser radiation is somewhat smaller. Information of this kind can be used to refine the details of the band structure of the crystal.

Surface melting under strong UV short laser pulse illumination can lead to surface amorphization as a result of extremely rapid solidification following laser-induced melting. Fig. 17 shows the orientational dependence of SHG

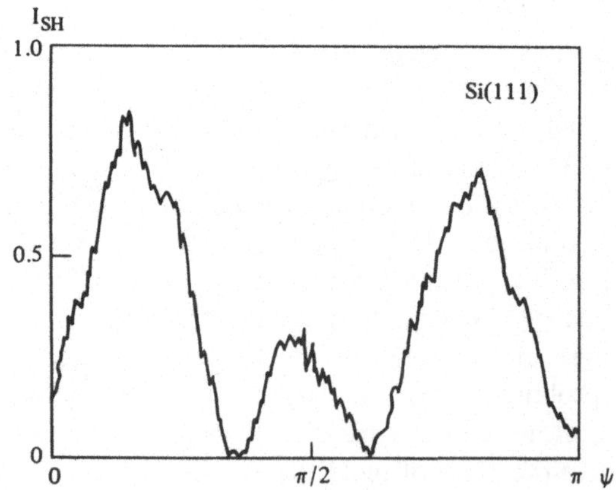

Fig. 17. Orientational dependence of SH from laser amorphized Si.

from amorphous film on Si surface. We attribute transformation of orientational dependence to amorphous material contribution and possibly to contribution of bulk Si influenced by thin amorphous film on its surface.

2.3.4. Subpicosecond dynamics of structural changes of Si surface in pulsed laser annealing

Shank et al. [126] used the technique of SHG in reflection for diagnostics of the fast (subpicosecond) initial phase of PLA of crystalline Si surface.
Reflection from (111) face was studied in these experiments, pumping and SH signal being polarized in the plane of the incidence. In the absence of annealing pulse as well as when the surface was probed by SHG prior to the arrival of annealing pulse (negative delays) the recorded dependence $I_{SH}(\psi)$ was well approximated by a theoretical curve of the form $I_{SH}(\psi) \sim (1 - \cos 3\psi)^2$ (see the angular dependence in Fig. 16a). Annealing was carried out by a single pulse with fluence $W > W_{th} = 0.2$ J/cm^2, wavelength $\lambda = 620$ nm and duration $\tau_p = 90$ fsec, incident normally on the (111) surface of crystalline Si. The probing was carried out by much more weak pulses of the same wavelength and duration, incident at an angle 45° to the irradiated section of the surface, with a precisely controllable delay after the annealing pulse. Fig. 18 b and c show the recorded dependences $I_{SH}(\psi)$ at time delays 0.3 and 1 ps after the annealing pulse. A drastic change of the character of this dependence clearly indicates the ultrafast change of the symmetry of surface, which is in a complete agreement with the idea of a laser-induced transition of the first order — the melting of the surface. The isotropic dependence $I_{SH}(\psi)$ with time delay 1 ps after

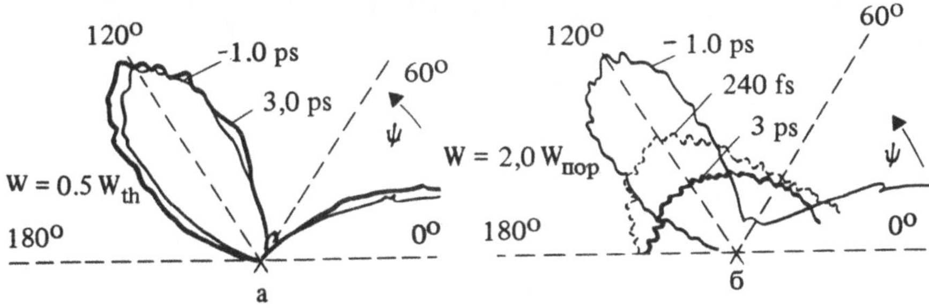

Fig. 18. Polar diagrams of the dependence of the intensity of the reflected second harmonic on the angle of rotation ψ of the crystal Si relative to the plane of incidence of the probing radiation for different values of the delay time between PLA and the SHG probing [126]. a) For $W = 0.5$ W_{th} (no melt); b) for $W \approx 2W_{th}$ (with $\tau_3 = 3$ ps the vanishing of the anisotropy of the diagram is seen clearly, indicating the appearance of a melt).

the annealing pulse corresponds to the isotropic melting of the surface layer.
The analogous results were received recently by Tom et al. [253] with the help
of 60 fs – pulses.

2.3.5. Nonlinear optical probing of stressed and deformed interfacial layers

Deformation of a crystal lattice caused by mechanical stress in the near surface
layers of multi-layered epitaxial structures and semiconductor-dielectric or
semiconductor-metal systems due to various lattice parameters or inter-atomic
distances in different materials at the interface is a difficult problem in semiconductor technology. This stress could be probed by optical means.
For example, as it was shown in [254], a SiO_2 film formation due to thermal
oxidation of silicon surface and the resulting appearance of strong mechanical
compression ($\sim$10 kbar) manifests itself in a dramatic increase (by a factor
of about 20) of SH intensity and a pronounced transformation of its orientational dependence.
The Si (111) sample with a thickness of about 0.3 mm were studied. Those
were oxidized in dry gaseous oxygen atmosphere at a temperature of 1100°C.
The oxide film had a thickness 500 Å. The experimental orientational dependence of SH signal is shown in Fig.19. In this case P-polarized SH component
was registered with P-polarized incident beam. Fig. 19a shows orientational
dependence of the SH signal from the Si surface with removed oxide layer
(howewer, it had a natural oxide film on it), and Fig. 19b shows orientational
dependence from the thermally oxidized Si.

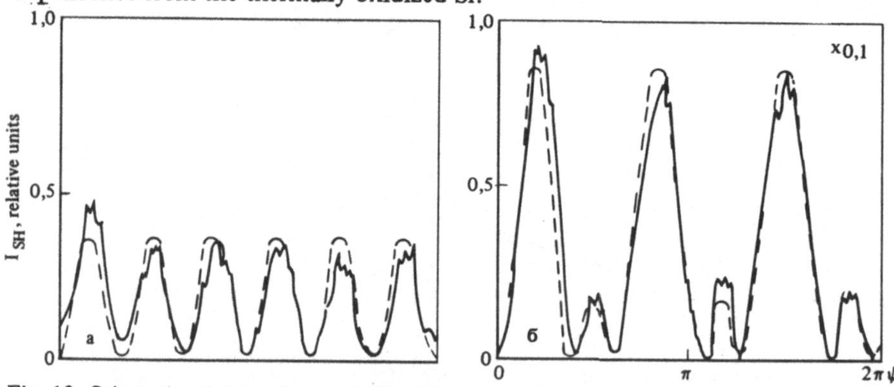

Fig. 19. Orientational dependence of SH. a) from pure Si (111); b) from Si (111)
with thermal oxide film (thickness 500 A).

We attribute the increase of SH intensity and its orientational dependence
transformation to lattice deformation in a subsurface layer, which leads to

the change of its symmetry, resulting in the appearance of strong electric dipole second order nonlinearity forbidden in the bulk material. Theoretical analysis of the induced dipole tensor $\chi_{def}^{(2)D}$ shows [255] that resultant orientational dependence of SH signal is similar to (2.29) but in this case isotropic contribution is expected to be comparable to anisotropic one ($|\zeta_Q/B| > 1$). The experiment shows that it is really the case.

Inhomogeneous mechanical stress in the near-surface layer is seen also in spontaneous Raman spectra. The observed spectral broadening (~ 5 cm^{-1}) enables one to estimate the peak value of mechanical stress (~ 10 kbar), since according to Anastasskis et al. [256] the pressure-induced Raman frequency shift is equal to 0.47 cm^{-1}/kbar. The authors of [26] observed similar effects from Si samples with silicide film deposited on its surface. SH intensity increased in this case by more 2 orders of magnitude.

The most striking effect due to the deformation − induced dipole contribution to SHG was observed with ion-implanted laser-annealed Si [254].

Fig. 20 shows the increase of SH intensity in ion-implanted laser-annealed Si on the implantation dose and type of implanted ions with respect to the SH

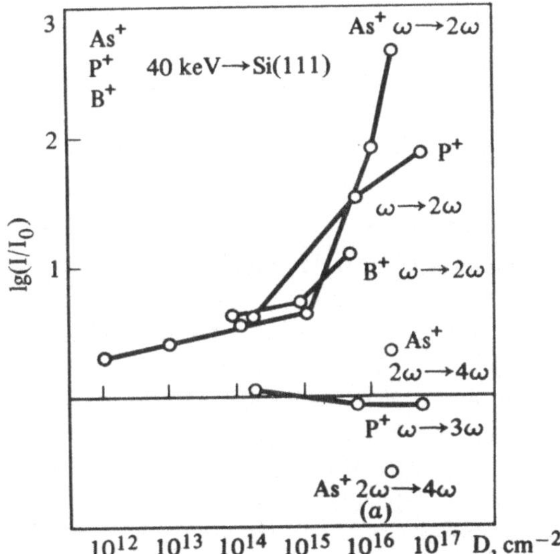

Fig. 20. Relative increase of SH intensity from ion-implanted pulsed-laser annealed Si.

signal from pure Si. With As$^+$ ion implantation (E=40 keV, D = $1.8 \cdot 10^{16}$ cm^{-2}) and subsequent pulsed ruby laser annealing SH intensity increased by nearly 3 orders of magnitude, when second harmonic from the fundamental Nd : YAG laser output (λ = 1.06 μm) was registered (process $\omega \quad - 2\omega$). However, no increase in SH intensity was observed when SH signal at λ = 0.26 μm was recorded (process $2\omega - \quad 4\omega$). The SH signal from ion-implanted Si without pulsed laser annealing did not show any increase either. Finally, there were no anomalous increase in SH intensity from ion-implanted and thermally annealed samples. Orientational dependence shown in Fig. 21 has only three strong peaks on a period compared with six ones from pure Si.

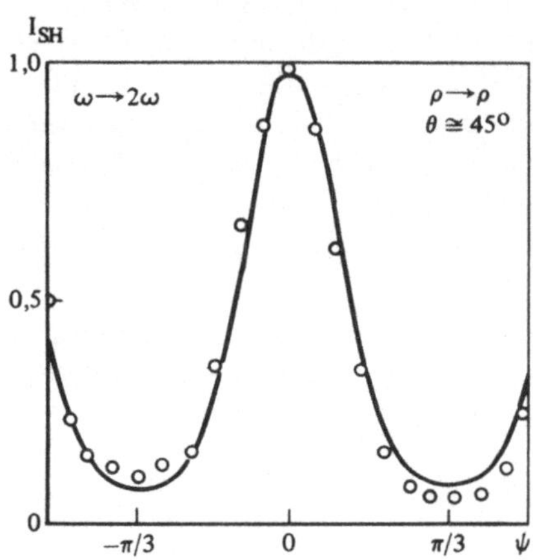

Fig. 21. Orientational dependence of SH signal from ion-implanted pulsed-laser annealed Si.

The observed peculiarities can be explained, if one assumes the appearance of the bulk electric dipole second order nonlinearity caused by inhomogeneous lattice deformation due to impurity ions. The theoretical analysis [255] shows that this inhomogeneous deformation must be proportional to the gradient of concentration of impurity ions having size and force constants different from that of host material. As it was stressed in Sec. 1, the ion-doped Si samples after nanosecond pulsed laser annealing can have impurity concentration exceeding equilibrium one. In addition, due to extremely short ($\sim 10^{-7}$s) melt duration, the depth distribution of impurities does not change appreciably. As a result, the maximum value of impurity concentration gradient is achieved right under the sample surface ($z \sim 1000$ A), while concentration gradient is nearly zero at the interface ($z = 0$). Thus, the strongest deformation of Si lattice causing the appearance of dipole contribution to $\chi_{def}^{(2)D}(2\omega)$ is expected at some depth inside the bulk. This explains why SH is strong in the case of $\omega - 2\omega$ process and does not show any increase in the case of $2\omega - 4\omega$. The reason is strong absorption of 4ω pulses so that very thin surface layer ($d \sim 100$ Å) is probed by $2\omega - 4\omega$ process, while in the case $\omega - 2\omega$ process the crystal thickness probed is an order of magnitude larger. Taking into account that impurity concentration gradient is nearly zero at the interface we conclude that there should be no dipole contribution to $\chi_{def}^{(2)D}$ in the case of $2\omega - 4\omega$ SH generation.

In the case of thermally annealed samples impurity ions can diffuse inside a crystal along relatively large distance during the annealing process (~ 30 min) thus leading to lower concentration gradient values and small dipole contribution to $\chi_{def}^{(2)D}$.

Note that SH intensity increase is not due to the decrease of fundamental bandgap in ion-implanted Si which could cause resonant increase of the bulk $\chi^{(2)Q}$, as well as $\chi^{(3)D}$. Fig. 20 shows that there is no increase of third harmonic intensity ($\omega - 3\omega$) though it must be expected if the resonant increase of nonlinear susceptibility really happened.

The orientational dependence of the SH signal (Fig. 21) is in agreement with the predicted structure of the dipole tensor induced by inhomogeneous uniaxial stress along the surface normal [255]. In fact, under the nonuniform deformation along the normal to the surface, the subsurface layer of (111)-Si no longer belongs to the symmetry class m3m but to class 3m; similarly the summetry of the deformed layer (100)-Si is lowered to 4m.

The analysis of the structure of the dipole tensor $\chi^{(2)D}$ in these classes leads to the orientational depencies of SH, presented in Table V.

The wavelength dependence of the SH signal shows also that silicon-impurity complexes do not play considerable role in SH increase since in this case the dipole contribution is expected to have the maximum value at the interface where impurity concentration is the largest. At the same time coherent and incoherent effects due to claster formation need further investigation.

2.3.6. The dynamics of Si lattice on a picosecond time-scale: picosecond CARS in reflection

Picosecond CARS investigation of Si under strong picosecond laser excitation is an demonstration of possibilities provided by nonlinear optical spectroscopy for the study of the lattice dynamic [256]. Computer simulation of the observed Raman spectra transformation enables one to get data on temperature, electron hole plasma concentration and amplitude of mechanical stress generated in Si via deformation potential mechanism.

Two RH6G dye lasers synchronously pumped by pulse train of the second harmonic of the passively mode-locked Nd : YAG laser output were used in [256] as coherent sources with frequencies ω_1 and ω_2 in traditional CARS

Table V.

The dependence of the intensity of the reflected SH on the angle
of rotation of the crystal around the normal to the surface ψ and
on the angle of incidence θ for Si due the presence of dipole quadratic
permeability $\chi_{ijk}^{(2)D}$, induced by uniaxial (normal to the surface) deformation

Crystal plane of SH reflection	Polarization of the incident wave	Polarization of the reflected harmonic	The dependence of $I_{SH}(\theta, \psi)$
(100)	p	s	0
	s	s	0
	s	p	$\lvert \chi_{311} \sin\theta \rvert^2$
	p	p	$\lvert \sin\theta ((-2\chi_{113} + \chi_{311}) \cos^2\theta + \chi_{333} \sin^2\theta) \rvert^2$
(111)	p	s	$\lvert \chi_{222} \cos^2\theta \sin 3\psi \rvert^2$
	s	s	$\lvert \chi_{222} \sin 3\psi \rvert^2$
	s	p	$\lvert \chi_{222} \cos\theta \cos 3\psi + \chi_{311} \sin\theta \rvert^2$
	p	p	$\lvert \chi_{222} \cos^3\theta \cos 3\psi + \cos\theta \sin\theta ((2\chi_{113} + \chi_{311}) \cos\theta + \chi_{333}) \rvert^2$

arrangement. One of the lasers (ω_2) was tuned in experiment to get anti-Stokes
spectrum at frequency $\omega_a = 2\omega_1 - \omega_2$ when the difference frequency $\omega_1 - \omega_2$
was scanned in the vicinity of the 520 cm^{-1} Raman resonance in Si. The wavelength of the ω_1 laser was fixed at the maximum of RH6G emission and its
output power was sufficiently strong to cause substantial heating — up to
the melting-of the sample surface. The microcomputer was used to accumulate
photon counts at each point of the CARS spectrum, to monitor laser energy
and to tune the dye lasers and the spectrometer.
The Si optical phonon spectra were recorded at various excitation laser pulse
fluencies E from E = (0.30 ... 0.05)E_0 to E = E_0, where E_0 = 0.2 J/cm^2 is
the threshold power density for surface melting; the laser illuminated spot on
the sample surface had the diameter of about 0.7 mm. At room temperature

without notable excitation the Si optical phonon spectrum is known to be rather narrow (3 cm^{-1}) strong Raman line with frequency of 520 cm^{-1}. As one can see from Fig. 22, the increase of the excitation laser pulse fluence induces dramatic change of both the line shape and the width as well as the decrease of the Raman line intensity compared with "nonresonant" electronic background. The Si lattice heating up is known to cause phonon mode "softening and line broadening and shift to the lower frequency side" [257]. On the contrary, in the experiment [256], the Raman spectrum remaines centered at 520 cm^{-1}. This feature is a result of mechanical stress generated in a sample in the presence of dense electron hole plasma leading to Raman frequency shift to higher frequency [256].

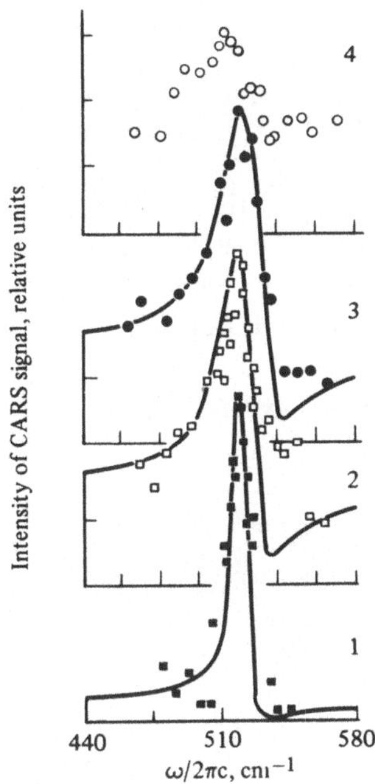

Fig. 22. Experimental (dots) and computer similated (curves) GARS spectra of laser excited Si for various laser fluences E(E$_0$ — threshold for melting).

Inhomogeneous nature of laser excitation is known to cause considerable distortion of spectral information obtained from the pulsed Raman spectra as it was discussed previously in Sec. 2.1. Thus, in order to get the quantitative estimate of temperature, concentration of photo-excited carriers n and amplitude of mechanical compression P in Si the computer simulation of the CARS spectra is needed. In so doing the authors of Ref. [256] took into account temperature dependence of optical parameters of Si, dependence of Raman line position and width on T and P and also spatial and temporal distribution of laser intensity and corresponding distribution of I. Thus, CARS spectra, both experimental and computer simulated, were essencially integrated ones.

In order to get data on temporal and spatial distribution of T, P and n we carried out numeric solution of the following set of basic equations:

$$\frac{\partial T}{\partial t} = I(t,z)(\hbar\omega_1 - E_g)(1-R)/d_0\,\rho\,c_p\,\hbar\omega_1\,,$$

$$\partial n(t,z)/\partial t = -\gamma_A n^3 + I(t,z)(1-R)/d_0\,\hbar\omega_1\,,$$

$$\frac{\partial^2 U}{\partial t^2} - c_1^2\,\frac{\partial^2 U}{\partial z^2} = \frac{\theta}{\rho}\,\frac{\partial n}{\partial z}\,.$$

Here U is atom displacement, E_g is the fundamental band gap, z is distance along the surface normal, n is density, R is reflectivity, γ_A is Auger recombination rate, θ deformation potential, d_0 depth of probing, c_p is a specific heat, $I(t,z) = I_0(t)e^{-z/d_0}$ laser intensity with Gaussian profils in time and in transverse plane.

An example of the calculated dependences n(t) and T(t) with $E = I(t)dt = 0.5\,E_0$ is shown in Fig. 23.

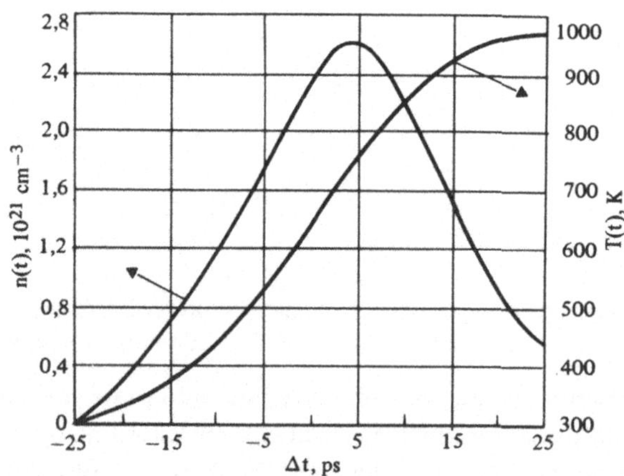

Fig. 23. The results of model calculations of time dependence of T, n, p of Si surface during picosecond pulsed ixcitation, $E = 0.5\,E_0$.

70

Deformation potential value θ and the depth of probing d_0 were varied in calculation in order to get best fit to the experimental spectra. The values obtained are $\theta = 15$ eV, $d_0 = 1000$ A.

Thus, the experimental results show that phonon mode is heated up during laser pulse. Computer simulation enabled to deduce data on n, T, P, characteristic for strongly excited Si. Application of subpicosecond laser pulses for CARS experiment might enable one to probe the dynamics of excitation in phonon branches.

So, nonlinear optical diagnostics of semiconductor surface compared with other methods of surface diagnostics has unique temporal resolution (till 10^{-14}s) enables to change the depth of probing by changing the probe pulse wavelength, has high spectral resolution, is able to probe interfaces between two dense media, does not need high vacuum and is relatively simple.

Many important problems could be studied using nonlinear optical diagnostics of semiconductors. Among them are processes of electron-phonon and phonon-phonon relaxation under strong laser excitation, mechanisms of nonlinear relaxation, lattice dynamics on a famtosecond time scale (for example, using CARS), nonlinear response of thin films (including superconducting films). A number of problems could be solved using combined methods, for example, combing nonlinear pico- and femtosecond spectroscopy with electron and X-ray spectroscopy.

3. NONLINEAR OPTICAL PROCESSES DETERMINED BY THE MODULATION OF THE SURFACE RELIEF AND OF THE SURFACE TEMPERATURE OF CONDENSED MEDIA

3.1. Laser-induced instabilities of the surface relief and the problem of the surface periodic structures (ripples) formation

3.1.1. Optical exitation of surface periodic structures

The problem of optically induced surface periodic structures (SPS) in the form of surface ripples formation has in recent years occupied an important place among the problems of the pulsed laser beam-condensed matter interactions. The typical experimental scheme of optical generation of SPS is very simple. The unfocused pulsed laser beam is incident on the absorping surface of the condensed matter. Inspite of the fact that the surface is under the action of almost plane wave, periodical modulation of the surface relief appears on the illuminated area. It is formed in the processes of interaction (its duration varies from 10^{-3} to 10^{-11} sec) and usually remains after the end of the laser pulse. Some examples of laser induced SPS are shown in Fig. 24. These structures were obtained on different materials by use of different lasers.

In should be noted, that a number of experiments, which were not limited solely to recording the "frozen" structures (the most efficient method is based on the study of the diffraction of the probe beam from a continuous laser [180]) (see below, Fig. 42) have been performed; making use of the technique of nonstationary diffraction of the probe beam, these experiments also followed the dynamics of the evolution of the structure during the action of the laser pulse (see, for example, [150—153, 202, 204]).

These experiments shows, that the time evolution of the SPS has the character of an instability; the dynamics of which have much in common with well known nonlinear optical instabilities in the stimulated scattering of light.

For the first time such SPS were observed in 1965 on the surfaces of Ge and Si semiconductors subjected to the action of the ruby laser pulses [135]. But intensive experimental and theoretical investigations of the effect of the SPS formation began approximately only in 1980. Among the first works in these field were [144, 104, 136, 137, 141].

The SPS formation is observed not only in semiconductors (Si, Ge, GaAs, InSb [135—143]), but also in the metals Ni, Gu, Pl, Al, steel, brass [144—146]), and in dielectrics (NaCl, fused and crystalline quartz [147—149, 267].

Besides the permanent surface ripples (left after the end of laser pulse) the reversible ripples, existing during laser pulse duration only are also observed

[150, 151]. Such reversible SPS are exited also under the action of the laser pulses on liquid metals [152] and melts of semiconductors [153].

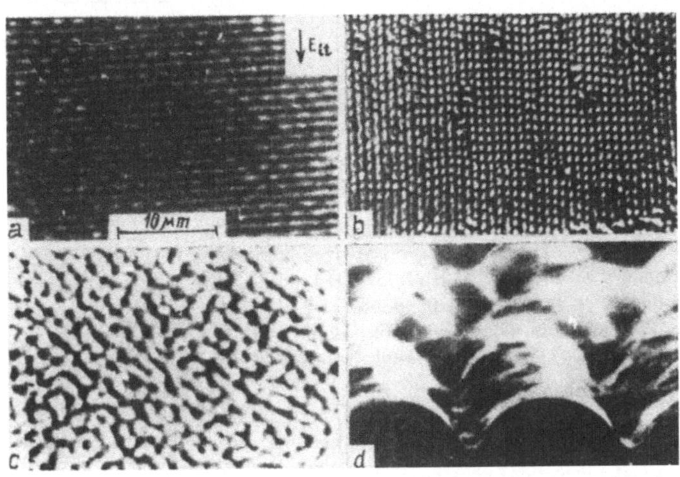

Fig. 24. Characteristic periodic structures induced by laser radiation on the surface of solids. a) One-dimensional grating, formed on the surface of a Ge semiconductor irradiated with a laser pulse with a wavelength $\lambda = 1.06$ μm ($I_i = 10$ MW/cm^2; $\tau_p = 100$ ns, the angle of incidence $\theta \approx 10^0$; the arrow marks the orientation of the vector E_{it} [180]; b) two-dimensional grating, formed on the surface of Ge under the action of high-intensity pulses of radiation with $\lambda = 1.06$ μm; the complicated structure could be linked with the interaction of gratings [141]; c) structure formed on the surface of germanium by a circularly polarized wave under oblique incidence (for the interpretation see Fig. 28) [180]; d) "frozen" capillary waves on the surface of quartz initially melted by a CO$_2$ laser (obtained with the help of a scanning electron microscope) [152].

Not only one-dimensional gratings, but also a two-dimensional structures [138] are observed, as well as a more complicated ordered formations [141] (see Fig. 24). The periods and orientations of the grating depend substantially on the laser light characteristics — the angle of incidence θ, polarization, frequency, and pulse energy (Fig. 25, 26).

The SPS formation occurs for a wide range of parameters of laser radiation ($\lambda = 0.38-10.6$ μm, pulse durations $\tau_p = 10$ps-1ms, and intensities $I_i = 5 \cdot 10^6 - 5 \cdot 10^8$ W/cm^2) (see Sec. 3.6).

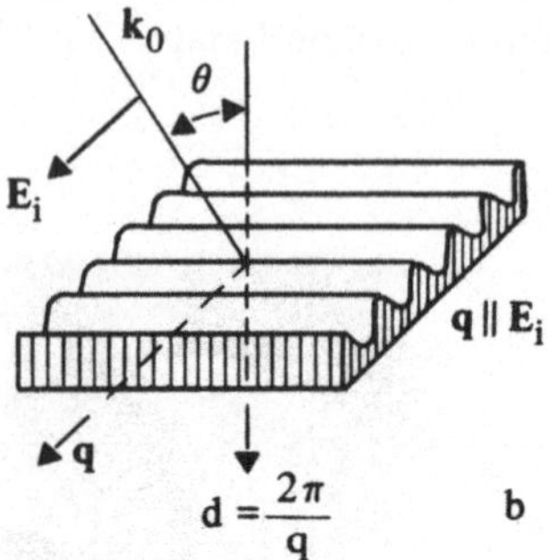

Fig. 25. Geometry of the formation of the most often encountered ("normal or longitudinal ($q \parallel E_i$)) gratings. a) p polarized pump wave (the vector E_i lies in the plane of incidence); b) s-polarized pump wave (the vector E_i is perpendicular to the plane of incidence); q is the wave vector of the wave of modulation of the relief, parallel to the projection of E_i on the plane of the surface; in both cases θ is the angle of incidence.

The experiments on optically induced SPS stimulated the intensive theoretical work [136, 137, 139, 199, 305, 165, 168—172, 220]. The physical picture of the generation of periodical structures has now been largely formulated.

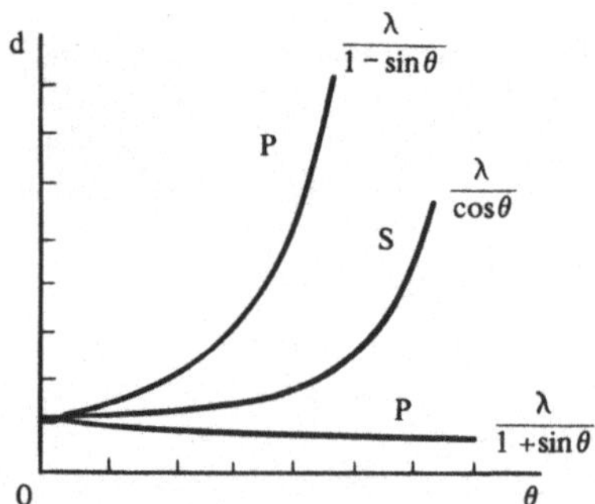

Fig. 26. Experimental dependence of the period of "normal" dominant surface gratings with s and p polarizations of the pumping wave as a function of the angle of incidence [137].

In the case of strongly absorbing surface the process of surface periodical structure formation can be represented schematically as follows.
1. Everything begins with the appearance of a periodically spatially modulated (interference) light field on the surface. This light field is formed by the interference of the incident wave with the wave scattered by the real nonuniform surface waves.

2. The light field with the periodically modulated intensity leads to the appearance of periodically modulated surface heating.

3. Finally, if the laser pulse intensity is high enough, then the nonuniform heating can give rise to the nonuniform melting and evaporation.

It is natural, that after the end of laser pulse the vestige of the nonuniform heating (and especially of the nonuniform melting and evaporation) creates a "frozen" SPS.

It is evident that these processes of the nonuniform heating, melting and evaporation are the stages of the complicated process of selfconsisted interaction between the laser light and the surface, undergoing the phase transition. The increase of SPS amplitudes affects the magnitude of the absorbed and difracted power so the feedback occurs. What is the sign of the this feedback? Are the positive feedback and hence the exponential grow of initial perturbations possible? What are the physical mechanisms of the feedback? These questions are, surely, the most interesting problems of the physics of SPS formation.

At present, one has the reasons in many cases to consider the positive feedback as the factor determining major features of SPS formation while the initial nonuniformitis play only the role of the "seed" for the exponentially growing process.

In the following, we, relying primarily on the work carried out in our laboratory [165, 168, 171, 172, 220], give the theoretical picture of the SPS formation, with the particular amphasis on the physics of laser-induced surface instabilities.

However, before considering the specific theoretical models, let us discuss in more details the qualitative picture of the phenomenon, which follows from the analisis of the experimental data.

3.1.2. The physics of rouph surface-laser field interaction

The first stage of the process is linked with the appearance of the interference light field on the surface. At this stage one may use the linear approximation and consider the diffraction of the incident laser wave with frequency ω on the Fourier-component of the surface relief with wave vector q, frequency Ω_q and amplitude ξ_q. As a result, one has two diffracted surface waves — the Stokes one with the frequency ω_1 and wave vector k_1 and anti-Stokes one (ω_{-1}, k_{-1}), the following conditions being satisfied:

$$\omega_1 = \omega - \Omega_q, \qquad \omega_{-1} = \omega + \Omega_q$$

$$k_1 = k_t - q, \qquad k_{-1} = k_t + q \tag{3.1a}$$

where k_t — is component of the incident wave vector on the plane $z=0$ (Fig. 27). The interference of the pump wave E_t, transmitted in the medium and of the two difracted waves E_a ($a = \pm 1$) creates the spatially and temporally periodical distribution of the light intensity in the subsurface layer.

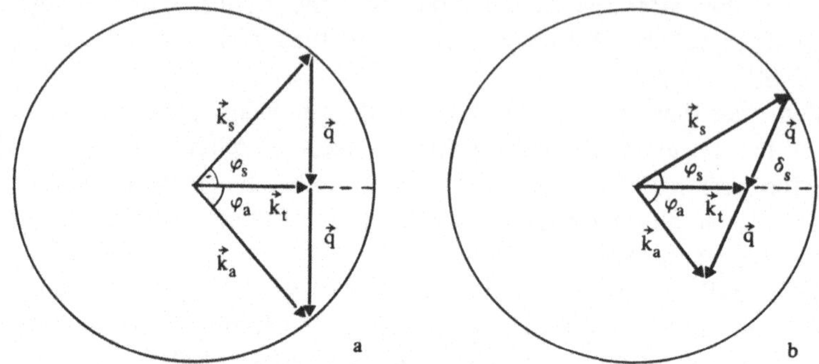

Fig. 27. Vector diagram of the conservation of momentum. a) Degenerate case of the relative orientation of the vector q and k_t (in this case $k_s = k_a \approx k_0$); b) nondegenerate case of the relative orientation of the vectors q and k_t (in this case $k_a \approx k_0$, $k_s \approx k_0$). $k_a \equiv k_1$, $k_a \equiv k_{-1}$.

The surface roughness is the most universial cause of the nonuniformity of real surface ($\Omega_q = 0$); besides in a number of cases, the surface acoustical waves (SAW) may play important role [155, 157], while the interaction with the melted surface substantially depends on the exitation of the capillary waves (CW) [158].

The conditions (3.1a) with $\Omega_q = 0$ enable constructing the rough picture of optically induced SPS on the metal surface with the dielectric permeability $\epsilon = \epsilon' + i\epsilon''$ (usually $|\epsilon'| \gg 1$) (see Fig. 25). It is well known, that in this case the surface electromagnetic wave (SEW) (exponentially decaying with the disfance from the surface) can propagate, the wave vector of which is

$$k_{surf} = \frac{\omega}{c} \left(\frac{|\epsilon'|}{|\epsilon'| - 1} \right)^{1/2} \approx \frac{\omega}{c} = k_0 \; .$$

Naturally, the incident p-polarized laser wave will exite a surface wave especially effectively, when $k_1 \approx k_{surf}$ or $k_{-1} \approx k_{surf}$

$$|k_t \pm q| \approx k_0 \tag{3.1b}$$

and the vectors k_1 ; k_{-1} lie in the plane of the surface.

From (3.1b), in the case of p-wave incident at an angle θ (see Fig. 25 a) one finds the period of "resonant" Fourie-component:

$$\frac{\omega}{c} \sin \theta \pm q = \pm \frac{\omega}{c} \left(\frac{|\epsilon'|}{|\epsilon'| - 1} \right)^{1/2} \approx \frac{\omega}{c} \, ,$$

whence

$$q = \frac{2\pi}{d} = \frac{\omega}{c} (1 \pm \sin \theta) \, , \qquad d = \frac{\lambda}{1 \pm \sin \theta} \, .$$

The quantity d determines the spatial period of the interference field, and, consequently, the period of optically induced SPS. The grating of this type ("longitudinal" gratings) are in fact observed in experiments. It is interesting to note that the condition analogous to (3.1b) has been considered already by Lord Rayleigh in connection with the so called Wood anomaly [200, 201]. The same resonances, in the case $\epsilon' < -1$ are, in the partically, responsible for the surface enhancement of Raman scattering (SERS) on the molecules absorbed on the sinusoidally modulated [159] or roughed [160] metal surface (see also review [161]).

3.1.3. The feedback mechanisms and evolution of structures

The considerations, presented above, are, of course, only guidelines. The SPS exitation mechanisms may differ for the different materials, different wavelengths and different laser radiation intensities. As an example of a situation, when the incident wave is diffracted not by static random relief, but rather by the propagating wave, we can mention the experiments, in which the laser beam intensity is so high, that prior to the SPS formation, the uniform surface melting occurs. In this case the SPS arise due to the capillary wave excitation (see [152] and experiment [202], where this was revealed especially clearly, as well as [215]).

The dispersion law for CW (these waves are formed owing to the surface tension in the liquid) has the form $\Omega_q = (\sigma q^3/\rho)^{1/2}$, where σ is coefficient of surface tension, ρ is the density of the liquid and the damping constant of CW is $\gamma_q = 2 \nu q^2$, where ν is the kinematic viscosity. Using for the Ge melt the value $\sigma \cong 6 \cdot 10^2$ g/cm^2, $\rho = 5$ g/cm^3, $\nu = 10^{-3}$ cm^2/s, as well as the fact that under laser exitation of CW $q \sim k_0 \sim 6 \cdot 10^4$ cm^{-1} ($\lambda = 1{,}06$ μm) (see Sec. 3.4), we obtain, that CW with $\Omega_q \sim 10^8$ c^{-1} and $\gamma_q \sim 10^7$ c^{-1} is exited, with the velocity of propagation $v_q = \Omega_q/q \sim 10^3$ cm/s.

As was already pointed out, the results of many experiments (see, for example, [202, 203]) indicate that the magnitudes of the amplitudes of starting surface roughness are not important for the laser induced SPS formation; this is a serious argument in support of the idea that positive feedback plays decisive role.

Below, we shall present the theory, describing all basic stages of the process of the SPS exitation. First the problem of diffraction of the incident laser wave E_i by the space-time Fourier component of the modulation of surface relief is solved. The general solution is given in Sec. 3.2; the expersions for the diffracted field E_1 and E_{-1} inside the medium are obtained in the approximation linear in ξ_q. (Analogous formula for the fields outside the medium describe the spontaneous Brilluon Scattering by SAW of CW [157]), nonlinear in ξ_q expressions are given in Sec. 3.6.1.

In the second step, the expressions obtained for E_1 and E_{-1} are used to calculate the temperature field. The last step — the closing of the feedback chain requires analysis of the equation for the specific surface exitation with the corresponding boundary conditions.

For SAW this is the equation for elastic displacement vector of the medium; for CW, these are the equations of hydrodynamics (3.20); and for the interference evoporation instability (IEI) this is equation for the velocity of the front of evaporation (3.17) (the last two equations must be treated simultaneously (see Sec. 3.4)).

It is important to note, that the solutions of the problem of diffraction (and, consequently, all the subsequent stage of the analysis also) are valid for media with an arbitrary values of dielectric permeability ϵ (ω). Thus, the results describing the exitation of SAW, CW or IEI are applicable for arbitrary condensed media — semiconductors, metals and dielectrics, as well as their melts.

In particular case $\epsilon' < -1$ (metal, the melts of semiconductors and polariton active dielectrics) the diffracted waves correspond to the induced SEW.

From the view point of the nonlinear optics the exitation of SAW or CW occurs as a result of stimulated scattering of light on the surface, analogous to the stimulated temperature scattering in the bulk (see review [162]). However, unlike bulk scattering, when the optical absorption is so low that the decay constant of scattered (surface) wave $\gamma_{\pm 1} \ll \gamma_q$ — the decay constants of the material exitation, the inverse inequality $\gamma_{\pm 1} \gg \gamma_q$ is realized on the surface. Thus, the scattered light wave (SEW) adiabatically adjusts to the acoustic wave (SAW or CW), which is amplified. The above mentioned analogy is manifested, for example, in the fact that, as in the case of the bulk stimulated temperature scattering, the critical intensity of SAW exitation is independent the linear optical absorption coefficient [165] (in spite of the fact, that the heating of the medium plays a primary role in these effects.).

It must be noted, that the stimulated scattering on the surface in the case of absorbing media, caused by the much more weak pondermotive forces, was studied earlier in [163], and in the case of transperent media in the work [164].
In this Section we shall present systematically the theory of the SPS — formation (the theory of instabilities of SAW, CW or IEI) in conclusion we shall discuss in detail the experimental data.

3.2. Diffraction of light waves by small spatial-temporal modulations of the surface relief

An arbitrary surface relief can be characterized by its space-time Fourier spectrum; thus it is sufficient in the linear case to consider the light wave diffraction by any one Fourier component.
Let the initially flat surface of the medium be the plane $z = 0$, and let the z-axis be directed into the medium. In the presence of modulation of the surface the medium fills the halfspace $z \geqslant \xi(r, t)$,

$$\xi(r,t) = \xi_q(t)\exp(-i\mathbf{q}\mathbf{r} + i\Omega_q t) + c.c., \tag{3.2}$$

where $r = \{x, y\}$ is a vector lying in the plane $z = 0$, $\xi_q(t)$ is the slow varying amplitude. Let a plane light wave is incident on the surface from vacuum

$$E(r, z, t) = E_i \exp(ik_t y + ik_z z - i\omega t) + c.c., \tag{3.3}$$

where k_z and k_t are the normal and tangential projections of the wave vector $\mathbf{k}_0$ (see Fig. 25).
Taking into acount only the first order diffraction, we write down the expressions for the field in vacuum ($z < \xi(r,t)$) as a superposition of the incident (i), reflected (r) waves with the frequency ω and two diffracted waves with the frequencies ω_a

$$E = (E_i e^{ik_z z} + E_r e^{-ik_z z}) \exp(ik_t y - i\omega t) +$$

$$+ \sum_{a=\pm 1} E_a' \exp(ik_a r + \Gamma_a z - i\omega_a t) + c.c., \tag{3.4}$$

where ω_a, k_1, k_{-1} are given by the equations (3.1a). The field inside the medium ($z \geqslant \xi(r,t)$) is the sum of the transmitted wave and two diffracted waves,

$$E = E_t \exp(ik_t y - \gamma z - i\omega t) +$$

$$+ \sum_{a=\pm 1} E_a \exp(ik_a r - \gamma_a z - i\omega_a t) + c.c.. \tag{3.5}$$

The problem of finding the expressions for E_r, E_z and E_a, E_a was considered in linear approximation in a number of works (see, for example, [159, 137]). For the x-rays spectral range analogous problem was considered in [206]. We shall use here the equations for the dynamical grating diffraction, obtained in [165], but written in more simple form, under the condition that $\Omega_q \ll \omega$. The nonlinear in ξ_q expressions for the amplitudes of the difracted field are obtained in [265] (see Sec. 3.6.1.) and used for the description of the nonlinear regime of capillar wave generation in [266] (Sec. 3.6.3.). The amplitudes of all fields in (3.4) and (3.5) are obtained from the condition that the tangential components of the vectors of the electric and magnetic fields be continuous at the interface $z = \xi\,(\mathbf{r}, t)$. Let us define the constants

$$\gamma^2 = k_t^2 - k_0^2\,\epsilon\,, \qquad \gamma_a^2 = k_a^2 - k_0^2\,\epsilon\,, \qquad \Gamma_a = k_a^2 - k_0^2\,,$$

$$(\mathrm{Re}\,\gamma, \gamma_a > 0\,, \qquad \mathrm{Re}\,\Gamma_a > 0\,, \qquad \text{for } k_a > k_0\,,$$

$$\mathrm{Im}\,\Gamma_a < 0 \qquad \text{for } k_a < k_0)\,. \tag{3.6}$$

Then the equations for the amplitudes of the field inside the medium can be written in the form

$$E_{tx} = \frac{2k_z}{k_z + i\gamma}\,E_{ix}\,, \qquad E_{ty} = \frac{-2i\gamma k_z}{k_t(k_z\epsilon + i\gamma)}\,E_{iz}\,,$$

$$E_{tx} = \frac{2k_z}{k_z\epsilon + i\gamma}\,E_{iz}\,, \tag{3.7}$$

$$E_{ax} = \frac{2k_z(1-\epsilon)}{\epsilon\Gamma_a + \gamma_a}\,\xi_a\left[\frac{k_{ay}^2 - \gamma_a\Gamma_a}{k_z + i\gamma}\,E_{ix} - ik_{ax}\frac{k_t\gamma_a - \gamma k_{ay}}{k_t(\epsilon k_z + i\gamma)}\,E_{ix}\right]\,, \tag{3.8}$$

$$E_{ay} = \frac{2k_z(1-\epsilon)}{\epsilon\Gamma_a + \gamma_a}\,\xi_a\left[\frac{-k_{ax}k_{ay}}{k_z + i\gamma}\,E_{ix} + i\,\frac{\gamma\Gamma_a\gamma_a - k_t\gamma_a k_{ay} - \gamma k_{ax}^2}{k_t(\epsilon k_z + i\gamma)}\,E_{iz}\right]\,, \tag{3.9}$$

$$E_{az} = \frac{2k_z(1-\epsilon)}{\epsilon\Gamma_a + \gamma_a}\,\xi_a\left[\frac{-ik_{ax}\Gamma_a}{k_z + i\gamma}\,E_{ix} + \frac{k_t k_a^2 - \gamma\Gamma_a k_{ay}}{k_t(\epsilon k_z + i\gamma)}\,E_{iz}\right]\,, \tag{3.10}$$

where $\epsilon \equiv \epsilon(\omega)$, $\xi_1 = \xi_q(t)$, $\xi_{-1} = \xi_q^*(t)$.

The equations (3.7–3.10) are valid for the media with the arbitrary values of $\epsilon(\omega)$. The analogous equations also exist for the fields outside the medium [172].

The diffracted field amplitudes (3.4–3.10) have a resonant dependence on k_a, owing to the presence of the quantity $\epsilon(\omega)\Gamma_a + \gamma_a$ in the denominators of (3.8)–(3.10). Let us consider, for example, the case of metals and the melts of the semiconductors, for which in optical region of the spectrum $\epsilon'(\omega) < 0$ and in addition

$$|\epsilon'(\omega)| \gg \epsilon''(\omega), \quad |\epsilon'(\omega)| \gg 1 \qquad (3.11)$$

In this case the resonance factor, in (3.8)–(3.10) is

$$l_a \equiv \frac{k_0(\epsilon-1)}{\epsilon(\omega_a)\Gamma_a + \gamma_a} \approx \frac{k_0}{|\epsilon'|^{1/2}(\Delta k_a - i\Gamma_p)} \; ; \qquad (3.12)$$

Here

$$\Delta k_a = k_a - \left(\frac{|\epsilon'|}{|\epsilon'|-1}\right)^{1/2} k_0 \; , \quad \Gamma_p = \frac{k_0\epsilon''}{2|\epsilon'|^2} \ll k_0 \; . \qquad (3.13)$$

The electromagnetic resonance width is determined by the SEW decay constant Γ_p.

From the equations (3.6), (3.8)–(3.10) one can see, that the diffracted waves amplitudes are resonancely enhanced (by factor $k_0/\Gamma_p \gg 1$) when the moduli of their wave vectors k_a coincide with that of free SEW (i.e. $\Delta k_a = 0$). Then the diffracted waves correspond to induced SEW (their amplitudes in accordance with (3.4)–(3.6) exponentially decay while away from the surface).

The diffracted waves resonance dependence on k_a takes place not only for metals and the melts of semiconductors ($\epsilon' < 0$), but also for dielectrics ($\epsilon' > 0$). This is illustrated by Fig. 28, which shows qualitative dependences of the resonance factor $l_a(a = s, a)$ on k_s in the case, when $|\epsilon| \gg 1$ ($\epsilon^{1/2} = n + im$). This resonance behavior of l_a is important for the interpretation of the experimental results on the generation of the SPS.

As follows from the expressions (3.8)–(3.10), the amplitudes of the diffracted waves grow linearly with ξ_q. In particular, for metals near the resonance we have from the formula (3.8) $E_{ax} \approx (2 \, k_0\xi_a)(|\epsilon'|/\epsilon'')E_{ix}$. From this expression it follows that already for grattings amplitudes of $\sim 100\text{Å}$, significant conversion of the pumping wave into the diffracted one should be expected. For example, in the case of copper ($\epsilon = -37 + i3,4; \lambda = 1,06 \, \mu m$) for a grating with an amplitude $2\xi_a = 100\text{Å}$ we have $E_{ax} = 0,6 \; E_{ix}$. As was shown in [137, 205] the nonlinear in ξ_q terms lead to the saturation of the growth of the amplitudes E_a and E_a' with ξ_q (see also Sec. 3.6.1).

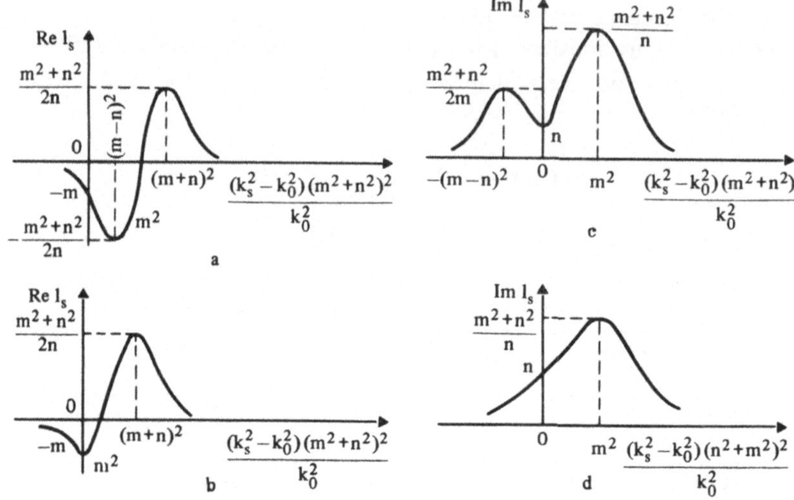

Fig. 28. a) Qualitative dependence of the real part of the electromagnetic factor Re l_s on k_s in the case of a metal $[\sqrt{\epsilon} = n + im, m>n, |\epsilon| \gg 1$ (according to the formula (3.85)]; b) qualitative dependence of Re l_s on k_s in the case of a dielectric ($m<n$, $|\epsilon| \gg 1$); c) qualitative dependence of the imaginary part of the electromagnetic factor Im l_s on k_s in the case of a metal [$m>n$, $|\epsilon| \gg 1$ (according to the formula (3.85))]; d) qualitative dependence of Im l_s on k_s in the case of a dielectric ($m<n$, $|\epsilon| \gg 1$).

Most theoretical works on SPS generation are limited to the calculation and analysis of the expression for the spatial harmonic of the power Q_q absorbed in the medium due to the interference of optical fields (electrodynamic model) [139, 136, 199, 207]. The expression for Q_q at surface ($z = 0$) has the form:

$$Q_q \equiv \left(E \frac{\partial D}{\partial t} \right)_{z=0} \sim (f_{+1} + f_{-1}) \exp(-iqx) + \text{c.c.},$$

where D — is the electric induction vector, f_1, f_{-1} are determined by the formula (3.58), (3.60), (3.61). From these expressions one can see, that Q_q is proportional to the electrodynamical resonance factors (see (3.12), (3.84), (3.85) and Fig. 28). According to the electrodynamical model, the extema of Q_q as a function of the variables q, φ_1 and φ_{-1} determine the periods and the orientations of the dominant gratings. The electrodynamical model enables predicting which gratings are possible, but it does not enable determining whether they are actually realized.

82

As it will be shown in Sec. 3.4. the growth rates of CW amplitudes also depend on resonance factors (3.84), (3.85) of formulae (3.86), (3.88) and, depending on the pumping fluency, the different situations can be realized, when one or another extremum of the resonance factors gives rise·to the formation of the dominant grating. Thus, the final conclusion about the realization of some specific grating and the optimal conditions for such a realization can be drawn only on the basis of the theory of the surface instabilities.

3.3. The feedback mechanisms in premilting regimes. The generation of coupled surface electromagnetic and acoustical waves

This effect arises due to nonuniform thermal expansion [165].
It is interesting primarily for the interpreting the experiments with reversible grating [150].

3.3.1. The exitation of the surface temperature wave owing to the interference of surface electromagnetic wave and transmitted laser wave

To describe the surface heating, one must add to the right hand side of heat conduction equation the term $(4\pi)^{-1}\,E\partial D/\partial t$ where E is determined by the equations (3.7–3.10) and the electric induction vector D has the form

$$D = \epsilon(\omega)\,E_t\,\exp\,(\,ik_t r - \gamma_z - i\omega t) +$$

$$+ \sum_{a=\pm 1} \epsilon(\omega_a)\,E_a\,\exp\,(ik_a r - \gamma_a z - i\omega_a t)\ . \tag{3.14}$$

The heat conduction equation in solid has the form

$$\frac{\partial T}{\partial t} = \chi\Delta T + \frac{1}{4\pi c_v}\,E\,\frac{\partial D}{\partial t}\ , \tag{3.15}$$

where $\chi = \kappa/c_v$, χ — is the temperature conductivity coefficient, c_v is the thermal capacity of the unit volume, κ is the thermal conductivity.
In the following we shall not be interested in uniform heating and consider only the surface temperature modulation in space and time owing to the product of amplitudes E_i and E_a' in the right hand side of (3.15). Using (3.14) we have in (3.15)

$$E \frac{\partial D}{\partial t} = \left\{ 2\omega\,\epsilon''(\omega) + \tfrac{1}{2}\Omega_q^2 \frac{\partial\epsilon''(\omega)}{\partial\omega} + i\,\omega\Omega_q \frac{\partial\epsilon'(\omega)}{\partial\omega} + \right.$$

$$\left. + i\Omega_q\,\epsilon'(\omega) \right\} \cdot \left\{ E_t^* E_s \exp\left[-(\gamma_1 + \gamma^*)z\right] + \right.$$

$$\left. + E_t E_{-1}^* \exp\left[-(\gamma_{-1}^* + \gamma)z\right] \right\} \exp(i\Omega_q t - i\mathbf{q}\mathbf{r}) + \text{c.c.}$$

We shall restrict ourselves with the case, when the following inequality holds $|\omega\epsilon''(\omega)| \gg |\Omega_q \epsilon'(\omega)|$. Assuming, further, the validity of conditions $\Omega_q \ll \omega$, $k_a \sim q \sim k_0 \ll \gamma_0$ and using the equations (3.7), (3.8)–(3.10), (3.14) we obtain

$$(4\pi c_v)^{-1} E\,\partial D/\partial t \approx a_q\,\xi_q(t)\exp\left(-i\mathbf{q}\mathbf{r} - \gamma_0 z + i\Omega_q t\right) + \text{k.c.}, \tag{3.16}$$

where in the case of s-polarized pump wave (vector $\mathbf{E}_i$ is perpendicular to the plane of incidence)

$$a_q = \frac{2\gamma_0\omega\,|E_i|^2 \cos^2\theta}{\pi c_v} \left\{ \frac{\Gamma_p \sin^2\varphi_1}{\Delta k_1 - i\Gamma_p} + \frac{\Gamma_p \sin^2\varphi_{-1}}{\Delta k_{-1} + i\Gamma_p} \right\}, \tag{3.17}$$

and the case of p-polarized pump wave (vector $\mathbf{E}_i$ in the plane of incidence)

$$a_q = \frac{2\gamma_0\,\omega\,|E_i|^2 \cos^2\theta}{\pi c_v(\cos^2\theta + |\epsilon|^{-1})} \left\{ \frac{\Gamma_p f_1}{\Delta k_1 - i\Gamma_p} + \frac{\Gamma_p f_{-1}}{\Delta k_{-1} + i\Gamma_p} \right\}, \tag{3.18}$$

Here $\gamma_a + \gamma^* \approx \gamma_0 = 2k_0|\epsilon'|^{1/2}$, $\cos\varphi_a = (\hat{k}_t \hat{k}_a)$, θ is the angle between $\mathbf{k}_0$ and the normal to the surface $z = 0$, i.e. the angle of incidence of the pump wave, Δk_a and Γ_p are defined in (3.13). $f_a = \cos^2\varphi_a - \sin\theta \cos\varphi_a$.

The thermal flux through the interface between the medium and the vacuum is equal to zero. Thus, up to the terms proportional to ξ_q^2, the initial and boundary conditions have the form

$$T(\mathbf{r}, z, t = 0) = T_0, \quad \left.\frac{\partial T\,(\mathbf{r}, z, t)}{\partial z}\right|_{z=0} = 0, \quad T(\mathbf{r}, z = \infty, T) = T_0. \tag{3.19}$$

For the times $t > \gamma_\pm^{-1}$ we put in (3.16)

$$\xi_q(t) = \xi_0 \exp(\gamma_q t) \tag{3.20}$$

84

and introduce the frequency

$$\Omega = \Omega_q - i\gamma_q \tag{3.21}$$

the temperature wave is installed on the times

$$t > t_T \equiv 1/2\, q^2\chi + (1 + z^2 q^2)^{1/2}/2q^2\chi \gg \gamma_\pm^{-1}.$$

Then the solution of the problem (3.15), (3.19), with taking into account (18), (22), can be written in the form

$$T(r,z,t) - T_0 \approx -\frac{\gamma_0}{\delta}\,\frac{a_q}{i\Omega - \chi\gamma_0^2}\,\xi_0 \exp(-i\mathbf{q}\mathbf{r} - \delta z + i\Omega t) + c.c.\,, \tag{3.22}$$

where $\delta^2 = q^2 + i\Omega/\chi$, $\mathrm{Re}\,\delta > 0$ and the quantity a_q is defined in (3.17),(3.18). The equation (3.22) is valid under the condition q, $|\delta| \ll \gamma_0$.

3.3.2. The laser generation of coherent surface acoustical waves and statical gratings

The temperature wave (3.22) gives rise to the driving force in the equation for the elastic displacement vector $\mathbf{U}$ [158]:

$$\rho\frac{\partial^2 \mathbf{U}}{\partial t^2} = \rho c_t^2\,\Delta\mathbf{U} + \rho(c_1^2 - c_t^2)\,\mathrm{grad}\,\mathrm{div}\,\mathbf{U} + \eta\Delta\dot{\mathbf{U}} +$$

$$+ (\eta/3 + \zeta)\,\mathrm{grad}\,\mathrm{div}\,\mathbf{U} - Ka\,\mathrm{grad}\,T\,, \tag{3.23}$$

where ρ is the density of the medium, c_1, c_t are the longitudinal and transverse sound velocities, η and ζ — the first and the second viscosity coefficients, K is the hydrostatic compression modulus, a — the thermal expansion coefficient.
On the free boundary in the linear approximation in ξ_q, the boundary conditions have the form

$$\frac{\partial U_x}{\partial z} + \frac{\partial U_z}{\partial x} = 0,\qquad \frac{\partial U_y}{\partial z} + \frac{\partial U_z}{\partial y} = 0, \tag{3.24}$$

$$-\frac{(T - T_{in})Ka}{\rho} + (c_1^2 - 2c_t^2)\left(\frac{\partial U_x}{\partial x} + \frac{\partial y_y}{\partial y}\right) + c_1^2\frac{\partial U_z}{\partial z} = 0$$

Let the x exis be directed along vector $\mathbf{q}$. Then the quantity $T - T_{in}$ in (3.22) does not depend on y. Thus $U_y = 0$, U_x and U_z do not depend on y, and the second of conditions (3.24) is automatically fullfilled.

The solution of the problem (3.22), (3.23), (3.24) has the form

$$U_x = \left(-\kappa_t A\, e^{-\kappa_t z} - q\, B\, e^{-\kappa_t z} + \right.$$

$$\left. + \frac{iq\, R\, \xi_0\, e^{-\delta z}}{\Omega^2 + c_l^2(\delta^2 - q^2) + i\Omega\eta'(\delta^2 - q^2)/\rho} \right) e^{-iqx + i\Omega t} + \text{c.c.} \qquad (3.25)$$

$$U_z = \left(iqA\, e^{-\kappa_t z} + i\kappa_l B\, e^{-\kappa_l z} + \right.$$

$$\left. + \frac{\delta\, R\, \xi_0\, e^{-\delta z}}{\Omega^2 + c_l^2(\delta^2 - q^2) + i\Omega\eta'(\delta^2 - q^2)/\rho} \right) e^{-iqx + i\Omega t} + \text{c.c.} \qquad (3.26)$$

Here

$$R = \frac{\gamma_0\, K\, a}{\rho\delta}\, \frac{a_q}{i\Omega - \chi\gamma_0^2}\,, \qquad \kappa_t^2 = q^2 - \frac{\Omega^2}{c_t^2 + i\Omega\eta/\rho}$$

and κ_l is obtained from κ_t by the substitution $c_t \to c_l$, $\eta \to \eta' = 4/3\,\eta + \zeta$,
A and B are some constants.
Since the surface relief modulation (3.2), (3,20) appear owing to SAW ($\xi(r,t) =$
$= U_z(r,z{=}0,t)$), from (3.2), (3.20) and (3.25) we have

$$\xi_0 = iqA + i\kappa_l B + \delta R\left[\Omega^2 + c_l^2(\delta^2 - q^2) + i\Omega\eta'(\delta^2 - q^2)/\rho\right]^{-1}\xi_0 \qquad (3.27)$$

Introducing now (3.25) in the boundary conditions (3.24), we obtain together
with (3.21) the system of three linear, homogenious algebraic equations for
the constants A, B and ξ_0. The solvability condition of this system (det $= 0$)
gives the dispersion equation, determining Ω. Neglecting the small terms, pro-
portional to $\eta^2 |E|^2$, we write down it in the form

$$c_l^2(q^2 - \kappa_l^2)(q^2 + \kappa_t^2) - 2c_t^2(q^2 + \kappa_t^2)q^2 + 4c_t^2 q^2 \kappa_l \kappa_t =$$

$$= \frac{c_t^2}{c_l^2}\, \frac{q^4 - \kappa_{0t}^4}{\delta + \kappa_{0t}}\cdot R\,, \qquad (3.28)$$

86

where

$$\kappa_{0l}^2 = q^2 - \Omega^2/c_l^2, \qquad \kappa_{0t}^2 = q^2 - \Omega^2/c_t^2. \tag{3.29}$$

Substituting (3.21), (3.26), (3.29) in (3.28) and equating the real and imaginary parts of this equation to zero, we obtain two equations, determining Ω_q and γ_q. In the absence of the external field (R = 0) and the dissipation ($\eta = \eta' = 0$) from (3.28) one obtains two different solutions.

1) The dynamical solution (SAW). Taking into acount the equation $c_l^2(q^2-\kappa_{0l}^2)=c_t^2(q^2 - \kappa_{0t}^2)$, which follows from (3.29) we reduce (3.28) to the following form

$$(\kappa_t^2 + q^2)^2 = 4q^2 \kappa_\perp \kappa_\parallel,$$

$$\kappa_\parallel^2 = q^2 - \Omega_q^2/c_l^2, \qquad \kappa_t^2 = q^2 - \Omega_q^2/c_t^2. \tag{3.30}$$

The solution of (3.30) is given by the usual formula for the frequency of Rayleigh SAW [155, 158]:

$$\Omega_q = q c_t \beta, \tag{3.31}$$

where β, as one can see from (3.30), is the function of c_t/c_l only. For the different solids $0{,}874 \leqslant \beta \leqslant 0{,}955$ [158].

2) The static solution: $\Omega_q = 0$, $\gamma_q = 0$ for any q.

Let us consider how these solutions are modified when $R \neq 0$ and $\eta \neq 0$, $\eta' \neq 0$. a) Let us substitude (3.21), (3.26), (3.29) into (9.28). Taking into acount that $|\gamma_q| \ll \Omega_q$, we linearize the equation (3.28) by γ_q, η, η'. The real part of the linearized equation gives again the equation (3.30), the solution of which is represented by the formula (3.31). Since Ω_q is big ($\Omega_q \gg \gamma_q$), we neglect the influence of pump field on the value of Ω_q. The imaginary part of linearized equation determines the formula for γ_q, which with taking into account (3.30), (3.31), (3.26) can be written in the form

$$\gamma_q = I / \sigma_0 - \gamma_\eta, \tag{3.32}$$

where

$$\sigma_0 = \frac{\kappa_\parallel}{\kappa_\perp} + \frac{\kappa_\perp c_t^2}{\kappa_\parallel c_l^2} - \frac{(q^2 + \kappa_\perp^2)}{q^2} \approx \frac{\kappa_\parallel}{\kappa_\perp} < 1 \tag{3.33}$$

87

and the SAW decay constant is

$$\gamma_\eta = \frac{\eta\beta^2 q^2}{4\sigma_0 q}\left(\frac{2\kappa_\parallel}{\kappa_\perp} - \frac{q^2+\kappa_\perp^2}{q^2}\right) +$$

$$+ \frac{\eta'\beta^2 q^2 c_t^2}{4\sigma_0\,\rho\,c_l^2}\left(\frac{2\kappa_\perp c_t^2}{\kappa_\parallel c_l^2} - \frac{q^2+\kappa_\perp^2}{q^2}\right) \approx \frac{\eta\beta^2 q^2}{2\rho}\ .$$

Further

$$I = - \frac{c_t^2}{4c_l^2}\,\frac{q^4-\kappa_\perp^4}{q^2\,\Omega_q}\,\frac{K a}{\rho\chi\gamma_0}\,\mathrm{Im}\,a_q\,[(1-i\Omega/\chi\gamma_0^2)\,\delta(\delta+\kappa_\parallel)]^{-1}\ , \qquad (3.34)$$

where a_q is defined in (3.17), (3.18) the growth rate of the amplitudes is expressed through the imaginary part of the product of two factors – one arising owing to SEW (which is proportional to a_q) and the second one, arising owing to the temperature wave.

Depending on the wavelength of the pump wave and the values of medium parameters one must distingwish two limiting cases. If $\chi\gamma_0^2 \gg |\Omega|$ and $\chi q \ll \beta c_t$, then in (3.34)

$$1 - \frac{i\Omega}{\chi\gamma_0^2}\,\delta\,(\delta+\kappa_\parallel) \approx \frac{i\Omega_q}{\chi} = \frac{i\beta c_t q}{\chi}$$

and the expression for the growth rate has the form

$$\gamma_q = \frac{K a\kappa_\perp}{4\beta^2 c_l^2\,\rho\gamma_0\kappa_\parallel}\,\mathrm{Re}\,a_q - \frac{\eta\beta^2 q^2}{2\rho} \qquad (3.35)$$

If $\chi\gamma_0^2 \gg |\Omega|$ and $\chi q \gg \beta c_t$, then

$$\gamma_q = - \frac{c_t^2}{c_l^2}\,\frac{K a\kappa_\perp}{8\Omega_q\rho\,\chi\gamma_0\kappa_\parallel}\,\mathrm{Im}\,a_q - \frac{\eta\beta^2 q^2}{2\rho}\ .$$

Let us estimate the critical intensity of SAW exitation. At a normal incidence in (3.1a) $k_t = 0$, then one obtains for the resonance SAW $q = k_a \sim k_0$ (see (3.13)). From (3.35) and (3.17) we have at $\sin^2\varphi_1 = \sin^2\varphi_{-1} = 1$

$$\max(\mathrm{Re}\,a_q) = 2\gamma_0\omega|E_i|^2/\pi c_v$$

Then from (3.6) and the condition $\gamma_q = 0$, we obtain the formula for the critical intensity

$$|E_i|^2_{th} = \gamma_\eta \frac{2\pi\beta^2 c_l^2 \rho c_v}{K a \omega} \left(\frac{1 - \beta^2 c_t^2 / c_l^2}{1 - \beta^2} \right)^{1/2} \tag{3.36}$$

For Cu one has $a = 7 \cdot 10^{-5} \mathrm{deg}^{-1}$, $K = 1{,}4 \cdot 10^{12} \frac{din}{cm^2}$, $c_v = 4 \cdot 10^7 erg^{-3}/cm^3 deg$, $\beta = 0{,}93$; $c_t = 2{,}3 \cdot 10^5$ cm sec^{-1}, $c_t^2/c_l^2 = 0{,}237$, $\rho = 9$ g/cm^3. For the pump wavelength $\lambda = 1\mu m$ ($\omega = 2 \cdot 10^{15}$ c^{-1}) with $\eta = 3{,}6 \cdot 10^{-8}$ din s/cm^2 we have from (3.33) $\gamma_\eta = 6{,}8 \cdot 10^6 s^{-1}$. Then from (3.36) for the threshold intensity of SAW exitation we obtain

$$I_{th} = \frac{c \, |E_i|^2_{th}}{2\pi} = 1{,}9 \cdot 10^7 \frac{W}{cm^2} \; .$$

b) Let us consider quasistatical case ($\Omega \approx 0$). We substitute (3.21), (3.26), (3.29) into (3.28) and put

$$|\Omega|/\chi q^2 , \quad |\Omega|/q c_t , \quad |\Omega|/\chi\gamma_0^2 , \quad \eta|\Omega|/\rho c_l^2 \ll 1 \; .$$

Then from (3.28) one obtains

$$\gamma_q = -\frac{2}{3}\left[\frac{Ka \, \mathrm{Re} \, a_q}{\gamma_0 \rho (c_l^2 - c_t^2)} + 2\chi q^2 \right], \quad \Omega_q = -\frac{2}{3} \frac{Ka \, \mathrm{Im} \, a_q}{\gamma_0 \rho (c_l^2 - c_t^2)} \; . \tag{3.37}$$

At a normal incidence $\mathrm{Im} \, a_q = 0$ and $\Omega_q = 0$. In (3.37), as one can see from (3.17) or (3.18), $\mathrm{Re} \, a_q < 0$ and $|\mathrm{Re} \, a_q|$ has maximum at $\Delta k_1 = \Delta k_{-1} = -\Gamma_p$. Then we obtain the soft relaxational mode

$$\gamma_q = \frac{2}{3}\left[\frac{2\omega Ka|E_i|^2}{\pi c_v \rho (c_l^2 - c_t^2)} - 2\chi q^2 \right] \; .$$

The critical intensity (at which $\gamma_q = 0$) for Cu at $\lambda = 10{,}6 \, \mu m$ is $I_{th} = 2 \cdot 10^8 W/cm^2$. At $I_i > I_{th}$ the sinusoidal statical grating, the period of which as it follows from (3.1a), where $k_t = 0$, is equal to the pump wavelength $\lambda = 2\pi/k_0$ and vector $\mathbf{q}$ is parallel to the vector $\mathbf{E}_i$, grows on the surface.

3.3.3. The Surface acoustical waves characteristics in dependance on the pump wave parameters

Let us restrict ourselves with the case (3.35), which is realized in solids. From (3.35) and (3.17), (3.18) one can see that those SAW have the maximal growth rate for which the resonance condition $\Delta k_a = \Gamma_p$ is fulfilled, i.e. according to (3.13)

$$k_a = (|\epsilon'| / |\epsilon'| - 1)^{1/2} (k_{0a} + \Gamma_p) \approx k_0 . \qquad (3.38)$$

Thus in (3.1a) the modul of vector k_a is fixed, but its the direction remains arbitrary. This means, that the ends of the possible vectors k_a are on the circle with radius k_0 (see Fig. 27). Then from (3.1a) one determines the modul and the direction of the wave vector of coupled SAW in dependence of SEW vector k_a orientation

$$q = k_0 (1 + \sin^2\theta - 2\sin\theta \cos\varphi_a)^{1/2}, \quad \cos\delta_a = k_0(\cos\varphi_a - \sin\theta)/q, \qquad (3.39)$$

where $\cos\varphi_a = (\hat{k}_t \hat{k}_a)$.

All possible cases of mutial orientations of vectors q and k_t are devided into two classes

a) Double-resonance of SEW: the vectors q and k_t are such, that resonance conditions (3.38) are fulfilled simultaneously for k_1 and k_{-1}. In this case in (3.17), (3.18), (3.35) the both terms with $a=1$ and $a=-1$ in the sum give equal contributions. The double resonance in the case of s-polarization of the pump and at sufficient small angles of incidence θ is realized at any orientation of q in respect to $k_t(|k_1| \sim |k_{-1}| \sim q)$. At sufficiently large θ it is realized only at orientation shown on Fig. 27. In this case $\cos\varphi_1 = \cos\varphi_{-1} = \sin\theta$. In the case of p-polarization the double resonance takes place only at sufficiently small angles θ (it is not realized at orientation shown on Fig. 27, because in this case $\cos\varphi_a = \sin\theta$, $f_a = 0$, so that $\gamma_q \leq 0$).

b) One-resonance case: the resonance conditions hold only for one of the vectors — either k_1 or k_{-1} (see Fig. 27). In this case in (3.17), (3.18), (3.35) the main contribution will give only one of the two terms; either with $a=1$, or $a=-1$. For s-polarization one resonance case is realized at $\theta \neq 0$ and $\cos\varphi_a \neq \sin\theta$. For p-polarization it takes place, when $\theta \neq 0$.

Let us consider firstly the case of s-polarization. Let the resonance condition (3.38) holds only for k_1. In this one-resonance case ($\theta \neq 0$ and $\cos\varphi_1 \neq \sin\theta$ according to (3.17) and (3.35) the growth rate γ_q may be written in the form

90

$$\frac{\gamma_q}{b_s} = -(\cos\varphi_s - \frac{a}{b_s}\sin\theta)^2 + (1-\frac{a}{b_s})(1-\frac{a}{b_s}\sin^2\theta)\,, \qquad (3.40)$$

where

$$b_s = \frac{Ka}{4\beta^2 c_1^2 \rho}\ \frac{\kappa_\perp}{\kappa_\parallel}\ \frac{\omega|E_i|^2}{\pi c_v}\cos^2\theta\,, \qquad a = \eta\,\frac{\beta^2 k_0^2}{2\rho}\,.$$

In the double resonance case, when $\theta \neq 0$ and $\cos\varphi_a = \sin\theta$ (Fig. 27) we have from (3.17), (3.35)

$$\gamma_q/b_s = 2(1-a/2b_s)(1-\sin^2\theta)\,. \qquad (3.41)$$

at small angle of incidence $\theta \approx 0$ at any $\cos\varphi_a = \sin\theta$ from (3.17), (3.35) we have

$$\gamma_q/b_1 = -2\cos^2\varphi_1 + 2(1-a/2b_1) \qquad (3.42)$$

The dependence γ_q/b_1 on $\cos\varphi_1$ according to the equations (3.40)–(3.42) is shown on Fig. 29a.

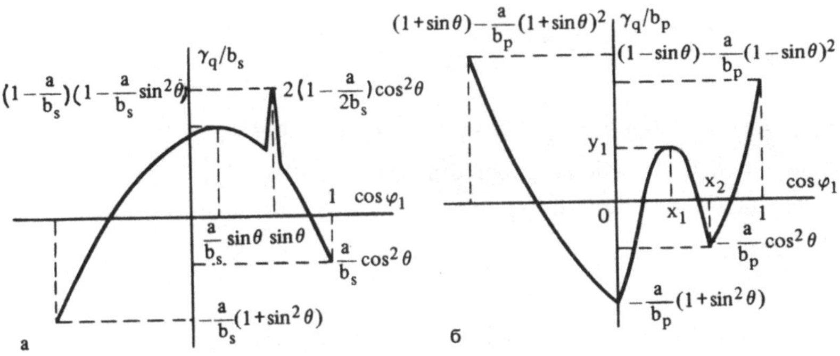

Fig. 29. a) The time-dependent growth rate γ_q of SAW as a function of $\cos\varphi_1$ [165] in the case of oblique incidence $(\theta \neq 0)$ of an s-polarized pump wave; b) dependence of the growth rate of SAW γ_q on $\cos\varphi_1$ in the case of oblique incidence $(\theta \neq 0)$ of a p-polarized pump. $x_1 = \sin\theta\,(1+2a/b_p)/2$, $x_2 = \sin\theta$, $y_1 = \sin^2\theta\,(1+4a^2/b_p^2)/4 - a/b_p$, $b_s \approx \dfrac{Ka}{4c_1^2}\dfrac{\omega|E_i|^2}{\pi c_v}\cos^2\theta$,

$$b_p = \frac{b_s}{\cos^2\theta + |\epsilon'|^{1/2}}\,, \qquad a \approx \eta\beta^2 k_0^2/2\rho\,.$$

The analisis of the equations (3.40)–(3.42) (see also Fig. 29a) permits to draw out the following conclusions:

1) The incidence of s-polarized pump wave at angle θ may lead to the generation ($\gamma_q > 0$) of the continuum of SAW the wave vectors of which are determined by the formula (3.39) and the frequencies determined by the formula (3.31). The necessary condition of the exitation of such SAW is the inequality $b_1 > a/2$.

2) If $\theta \neq 0$ and

$$2 \cos^2 \theta \, (1 - \frac{a}{2b_s}) > (1 - \frac{a}{b_s}) \, (1 - \frac{a}{b_s} \sin^2 \theta)$$

(see Fig. 29a), then the absolute maximum of the growth rate γ_q at given θ is attained at such orientation of vector q that $\cos\varphi_1 = \cos\varphi_{-1} = \sin\theta$. In this case SAW with wave vector q, which, according to (3.39), is determined by formula

$$q \perp k_t \quad (\text{i.e. } q \parallel E_i) ; \quad q = k_0 \cos \theta \tag{3.43}$$

will be generated.

3) If

$$(1 - \frac{a}{b_s}) \, (1 - \frac{a}{b_s} \sin\theta) > 2(1 - \frac{a}{2b_s}) \cos^2 \theta ; \quad b_s > a$$

(see Fig. 29a) the absolute maximum of γ_q is at $\cos \varphi_1 = \frac{a}{b_s} \sin \theta$ and most effectively two SAW will be generated with wave vectors q, the directions of which are determined by the angles δ and $2\pi - \delta$ (the angle δ is determined on Fig. 27) and

$$q = k_0 \left[1 + (1 - \frac{2a}{b_s}) \sin^2 \theta \right]^{1/2} , \quad \cos\delta = \frac{k_0 (1 - a/b_s) \sin\theta}{q} . \tag{3.44}$$

4) At $\theta = 0$ the continuum of SAW will be generated with the growth rates (3.42). The SAW with the parameters almost such as the parameters of SAW with $\cos\varphi_1 = 0$ will be generated most effectivly; for them: $q \approx k_0, q \parallel E_i$.

Let us consider now the case of p-polarization of pump wave. From (3.18), (3.35) one can see that at $f_a > 0$ the growth rate has a maximum for SAW with $\Delta k_a = \Gamma_p$ and at $f_a < 0$ — for SAW with $\Delta k_a = -\Gamma_p$. From these resonance conditions it follows, that for these SAW, as in (3.38), $k_a \cong k_0$. Let up assumè further that the rescnance condition is fulfilled only for $a = 1$.

In one resonance case ($\theta \neq 0$) for growth rate γ_q, from (3.20), (3.35) one has

$$\frac{\gamma_q}{b_p} = |\cos^2 \varphi_1 - \sin\theta \, \cos\varphi_1| - \frac{a}{b_p}(1 + \sin^2\theta - 2\sin\theta \, \cos\varphi_1) ,$$

$$b_p = \frac{K\,a}{4\beta^2 c_l^2 \rho} \frac{\kappa_\perp}{\kappa_\parallel} \frac{\omega |E_i|^2 \cos^2\theta}{\pi c_V (\cos^2\theta + |\epsilon'|^{-1})}$$

$$(3.45)$$

and the quantaty a is defined in (3.40).

The dependancies γ_q – (3.45) on $\cos\varphi_1$ in the case of sufficiently large pump intensities ($b_p > 2a$) are shown on Fig. 29b. The magnitude and the direction of $\mathbf{q}$ for the corresponding SAW are given by (3.39).

In the double resonance case ($\theta = 0$, $\cos\varphi_1 = -\cos\varphi_{-1}$, $k_1 = k_{-1}$) the growth rate γ_q has the form

$$\gamma_q / b_p = 2\cos^2 \varphi_1 - a/b_p . \qquad (3.46)$$

The analisis of the formulas (3.45), (3.46), and also Fig. 29b, permits one to draw out the following conclusions:

1) At intensities $b_p > a(1 - \sin\theta)$ the SAW are generated with $\cos\varphi_1 = 1$ with vector $\mathbf{q}$ directed antiparallel to $\mathbf{k}_t$:

$$\mathbf{q} \uparrow\downarrow \mathbf{k}_t , \qquad q = k_0(1 - \sin\theta) . \qquad (3.47)$$

With the intensity growth ($b_p > a(1 + \sin\theta)$) SAW with $\mathbf{q}$ parallel to $\mathbf{k}_t$ is generated additionally:

$$\mathbf{q} \uparrow\uparrow \mathbf{k}_t , \qquad q = k_0(1 + \sin\theta) . \qquad (3.48)$$

2) At big intensities ($b_p \gg a$) the growth rate of the SAW – (3.48) is greater then that of SAW – (3.47).

We have considered the process of selfexitation of running SAW with the wave vector $\mathbf{q}$, defined in the form (3.2). Analogous consideration shows that for SAW with wave vector $-\mathbf{q}$ the growth rate $\gamma_{-q} = \gamma_q$. This means that two oppositevly running SAW are generated simaltaneously, the superposition of which may form a standing SAW.

3.4. The capillary waves and evaporation waves generation
under the action of laser radiation on the liquid metals, semiconductors
and dielectrics

In the presence of a spatially uniform surface melt (the pump energy density W exceeds the melting threshold: $W > W_m$) two instabilites play the basic role: the capillary wave (CW) instability [146, 168, 169] and the interference evaporation instability (IEI) [170–173]. Usually these processes are considered separately. However, under certain conditions the hydrodynamic and evaporation effects may influence one another [169, 174]. In this section we consider both these instabilities on the basis of unified treatment [220].

Three mechanisms of feedback, accompanying CW generation or IEI, may operate depending on the degree of heating of the surface: 1) the thermo-capillar forces, which arise due to the surface tension (σ) dependence of temperature T, build up CW with the frequency Ω_q and wave vector $\mathbf{q}$; 2) the vapor recoil pressure, arising in the process of spatially nonuniform evaporation of matter, makes an additional contribution to building up of CW; 3) finally, the process of direct mass outflow, accompaning spatially nonuniform evaporation, also increases the amplitudes of the modulation of the relief.

In this section the general dispersion equation taking into account the contributions of both the evaporative mechanism and generation of CW is obtained. The solution of this equation, $\Omega = \Omega_q - i\gamma_q$ determines the growth rate γ_q and the frequency Ω_q in dependence of q, ω, θ, the intensity and the polarization of the pumping wave for media with arbitrary values of the dielectric permeability $\epsilon = \epsilon' + i\epsilon'' = (n + im)^2$. This dispersion equation enables deter᛫mining the regions of domination of each of three above mentioned SPS – generation mechanisms and describing from the unified point of view the generation of both the longitudinal gratings ($\mathbf{q} \parallel \mathbf{E}_{it}$, $q \sim k_0$, see Fig. 25) and transverse ones ($\mathbf{q} \perp \mathbf{E}_{it}$) as well as smallscale ($q \gg k_0$) structures ($\mathbf{q} \perp \mathbf{E}_{it}$), observed in the works [146, 175] ($\mathbf{E}_t$ is the projection of $\mathbf{E}_i$ on the plane $z=0$).

It is important to note, that the dispersion law of CW, responsible for the SPS-generation, depends substancially on the laser radiation parameters. Thus, the CW, considered here, form in fact essentially new class of CW, which can be called laser-induced CW. The CW dispersion law, obtained here, describes the effect of the generation of CW doublets (see Sec. 3.4.5). From the general dispersion equation the existence of laser-induced evaporation waves also follows, the dispersional characteristics of which depend on the laser pump parameters and the value of $\epsilon(\omega)$ (see Sec. 3.5).

94

3.4.1. Spatially nonuniform heating, evaporation and motion of liquid under the action of laser radiation

We assume that the melt occupies the half space $z > \xi(x, t)$

$$\xi(x, t) = \xi_1(x, t) + \xi_2(x, t) = [\xi_1(t) + \xi_2(t)]\, e^{-iqx} + c.c. =$$

$$= \xi_q(t)\, e^{-iqx} + c.c. \tag{3.49}$$

where $\xi_1(x,t)$ is an arbitrary Fourier component of surface relief, formed when matter is revomed from the surface during the evaporation prosses, and $\xi_2(x,t)$ is the relief, arising owing to the CW exitation (note, that in (3.49) in contrast to (3.2) $\xi_q(t)$ contains the full time dependence). We take into account, that due to the matter evaporation from the surface, the interface (3.49) moves as a whole with a velocity v_0 along the z-axis. Let us introduce the moving coordinate system x, y, z, related to the stationary coordinate by the relations $x = x'$, $y = y'$, $z = z' - v_0 t$. In the moving coordinate system the interface is described by the formula (3.49) and the equations for the fields (3.5) do not change owing to the shortness of the time over which the stationary values of amplitudes E_t, E_a are established (in calculating these quantities it may be assumed that $\xi_q(t) = const$, $v_0 = 0$). The saturation vapor pressure over the surface is given by the formula [176, 177]:

$$p_{sat} = p_c \exp\left[\frac{U}{R}\left(\frac{1}{T_c} - \frac{1}{T(x,z,t)}\right) - \frac{\sigma}{RT(x,z,t)}\, q^2 \xi(x,t)\right], \tag{3.50}$$

the temperature dependence of the rate of evaporation of the matter is given by equation

$$v_0 + \frac{\partial \xi_1(x,t)}{\partial t} = c_0 \exp\left[-\frac{U}{RT(x,z,t)} - \frac{\sigma q^2}{RT(x,z,t)}\, q^2 \xi(x,t)\right], \tag{3.51}$$

where p_c is the saturated vapor pressure at the surface temperature T_c, U is the activation energy per unit volume, $R = \bar{n}k_B$, $\bar{n}$ is the number of molecules of the material per unit volume, k_B is Boltzman's constant. The constant c_0 is determined from the experimental dependence of p_{sat} on T, if the following equation is used [176]:

$$p_{sat} = v_0 (2\pi \rho RT)^{1/2} \tag{3.52}$$

where ρ is the matter density. If one assumes, that in the process of evaporation the molecules returning back to the surface stick to it, then the recoil vapour pressure is approximately

$$p \approx \frac{p_{sat}}{2} \tag{3.53}$$

We assume that the melt is an uncompressible liquid. In the moving coordinate system we have the following hydrodynamic equations [158]:

$$\Delta \varphi = 0, \quad \frac{\partial \mathbf{A}}{\partial t} - v_0 \frac{\partial \mathbf{A}}{\partial z} = \nu \Delta \mathbf{A}, \quad \operatorname{div} \mathbf{A} = 0,$$

$$\mathbf{v} = \operatorname{grad} \varphi + \operatorname{rot} \mathbf{A} \tag{3.54}$$

where ν is kinematic viscosity coefficient, φ, $\mathbf{A}$ are the scalar and vector potentials of liquid velocity $\mathbf{v}$. We represent the temperature in the form

$$T(x, z, t) = T_0(z) + T_1(x, z, t), \tag{3.55}$$

where $T_0(z)$ is the temperature, characterizing the uniform heating along the surface $z = 0$, and $T_1(x, z, t) \sim \xi(x, t)$ describes spatially nonuniform heating of the liquid. Up to the terms, proportional to ξ^2, the boundary conditions at $z = 0$ may be written, with taking into account of (3.50) – (3.53), in the following form

$$\frac{\partial \varphi}{\partial t} - v_0 \frac{\partial \varphi}{\partial z} - g_0 \xi_1(x,t) + \sigma_0 \frac{\partial^2 \xi(x,t)}{\partial x^2} + 2\nu \frac{\partial v_z}{\partial z} + p_{0T}\left(T_1 + \frac{\partial T_0}{\partial z}\xi(x,t)\right) -$$

$$- \frac{p_0 \sigma q^2}{RT_{0s}} \xi(x,t) = 0$$

$$\nu\left(\frac{\partial v_x}{\partial z} + \frac{\partial v_z}{\partial x}\right) + \sigma_{0T}\left(\frac{\partial T_1}{\partial x} + \frac{\partial T_0}{\partial z}\frac{\partial \xi(x,t)}{\partial x}\right) = 0, \tag{3.56}$$

where $p_0 = p_{in}/2\rho$, $\sigma_0 = \sigma/\rho$, $\sigma_{0T} = d\sigma/\rho dT$, $p_{0T} = dp_{in}/2\rho dT$, $T_{0s} = T_0(z = 0)$, $g_0 = g_c/\rho$ and g_c is the acceleration of gravity. The velocity component v_z is related to $\xi_2(x, t)$ by the relation

$$v_z(x, z=0, t) = \frac{\partial \xi_2(t)}{\partial t} \tag{3.57}$$

In the moving coordinate system, the heating of the liquid in the presence of evaporation from a surface with the use of (3.7) — (3.10), is described by the equation

$$\frac{\partial T}{\partial t} - v_0 \frac{\partial T}{\partial z} + \mathbf{v}\nabla T = \chi \Delta T + f_0 e^{-\gamma_0 z} + [(f_1 e^{-(\gamma_1 + \gamma^*)} +$$

$$+ f_{-1} e^{-(\gamma^*_{-1} + \gamma)z}) \xi_q(t) e^{-iqx} + c.c.],$$

$$T(x, z, t=0) = T_{in}, \qquad T(x, z=\infty, t) = T_{in}, \tag{3.58}$$

$$\left. \frac{\partial T}{\partial n} \right|_{z=\xi(x,t)} = \frac{g}{\kappa} \left(v_0 + \frac{\partial \xi_1(x,t)}{\partial t} \right), \tag{3.59}$$

where $\kappa = c_v \chi$, c_v is the heat capacity per unit volume, $\gamma_0 = \gamma + \gamma^*$ and g is the latent heat of vaporization per unit volume, T_{in} is the initial temperature and $\mathbf{n}$ is the unit vector normal to the surface $z = \xi(x, t)$. The quantaties f_0, f_1, f_{-1}, characterizing the intensity of the heat sources in the case of s-polirized incident wave and with arbitrary $\epsilon(\omega)$ are given by

$$f_0 = 2\omega \epsilon'' k_z^2 |E_i|^2 / \pi c_v |k_z + i\gamma|^2 ,$$

$$f_1 = f_0(\epsilon - 1) \left(\frac{\gamma_1 \Gamma_1 \sin^2 \varphi_1}{\epsilon \Gamma_1 + \gamma_1} - \frac{k_0^2 \cos^2 \varphi_1}{\Gamma_1 + \gamma_1} \right); \tag{3.60}$$

and in the case of p-polarized pumping wave we have

$$f_0 = (2\omega \epsilon'' / \pi c_v)(k_z^2 / |\epsilon k_z + i\gamma|^2)[(|\gamma|^2 + k_t^2)/k_0^2] |E_i|^2 ,$$

$$f_1 = \frac{2\omega \epsilon''}{\pi c_v} \frac{k_z^2}{|\epsilon k_z + i\gamma|^2} |E_i|^2 (1-\epsilon) \cdot$$

$$\cdot \frac{|\gamma|^2 (k_t^2 \sin^2 \varphi_1 - \gamma_1 \Gamma_1) + k_1 k_t [k_t k_1 + (\gamma_1 \gamma^* - \gamma \Gamma_1)\cos\varphi_1]}{k_0^2 (\epsilon \Gamma_1 + \gamma_1)} \tag{3.61}$$

where $\cos \varphi_a = (k_t k_a)/k_t k$, $a = \pm 1$. The quantities f_{-1} for s or p-polarized pumping waves are obtained from (3.60) (3.61) by the substitution: $\epsilon \to \epsilon^*$, $k_1 \to k_{-1}, \varphi_1 \to \varphi_{-1}, \gamma_1 \to \gamma_{-1}, \Gamma_1 \to \Gamma_{-1}$, we shall solve the problem

(3.58)–(3.61) in two stages. In the stationary state, in the zeroth-order approximation in $\xi(x, t)$ we have from (3.58)

$$- v_0 \frac{\partial T_0(z)}{\partial z} = \chi \Delta T_{0s}(z) + f_0 e^{-\gamma_0 z}$$

$$T_{0s}(z=0) = T_{0n} , \quad T_0(z=\infty) = T_{in}$$

$$\left. \frac{dT_0(z)}{dz} \right|_{z=0} = \frac{g v_0}{\kappa} , \tag{3.62}$$

where T_{0s} is the as yet unknown surface temperature. The stationary solution of the problem (3.62) can be written in the form

$$T_0(z) = A e^{-\gamma_0 z} + B e^{-v_0 z/\chi} + T_{in} ,$$

$$A = - \frac{f_0}{\gamma_0(\gamma_0 \chi - v_0)} , \qquad B = \frac{f_0 \chi}{v_0(\chi \gamma_0 - v_0)} - \frac{g}{c_v} , \tag{3.63}$$

$$v_0 = \frac{f_0 c_v}{\gamma_0 [g + c_v(T_{0s} - T_{in})]} , \quad (A + B = T_{0s} - T_{in}) . \tag{3.64}$$

In the zeroth-order in $\xi(x, t)$ approxiation, from (3.51´) one gets

$$v_0 = c_0 e^{-v/R T_{0s}} \tag{3.65}$$

The equations (3.64), (3.65) determine the stationary values v_0 and T_{0s} as the functions of the laser beam polarization and fluence.

Now we calculate the spatially nonuniform (along the surface $z = 0$) temperature distribution $T_1(x, z, t)$. In the linear in $\xi(x, t)$ approximation, from (3.58), (3.59), taking into account (3.62), we have the following set of equations

$$\frac{\partial T_1}{\partial t} - v_0 \frac{\partial T_1}{\partial z} = \chi \Delta T_1 - v_z \frac{\partial T_0}{\partial z} + [(f_1 e^{-(\gamma_1 + \gamma^*)z} +$$

$$+ f_{-1} e^{-(\gamma^*_{-1} + \gamma)z}) \xi_q(t) e^{-iqx} + c.c.]$$

$$T_1(x, z, t=0) = 0 , \quad T_1(x, z=\infty, t) = 0,$$

$$\left.\frac{\partial T_1(x,z,t)}{\partial z}\right|_{z=0} + \left.\frac{\partial^2 T_0(z)}{\partial z^2}\right|_{z=0} \xi(x,t) = \frac{g}{\kappa}\frac{\partial \xi_1(x,t)}{\partial t}. \qquad (3.66)$$

The solve these equations it is necessary to calculate v_z. The stationary solution of equation (3.54) has the form

$$\varphi = \varphi_0 \exp(-iqx - qz + i\Omega t) + \text{c.c.}, \quad \Omega = \Omega_q - i\gamma_q ,$$

$$A_y = a_0 \exp(-iqx - \delta z + i\Omega t) + \text{c.c.}, \quad A_x = A_z = 0 , \qquad (3.67)$$

$$\xi_1(t) = \xi_{10}\exp(i\Omega t), \quad \xi_2(t) = \xi_{20}\exp(i\Omega t), \quad \xi_0 = \xi_{10} + \xi_{20} , \qquad (3.68)$$

where φ_0, a_0, ξ_0 are the fluctuation initial values, δ is some constant, substituting (3.61) into (3.54), we have

$$i\Omega + v_0\delta = \nu(\delta^2 - q^2), \quad \text{i.e. } \delta = \frac{v_0}{2\nu} + \left(q^2 + \frac{i\Omega}{\nu} + \frac{v_0^2}{4\nu^2}\right)^{1/2} \qquad (3.69)$$

$$v_x = (-iq\varphi_0 e^{-qz} + \delta a_0 e^{-\delta z}) e^{-iqx+i\Omega t} + \text{c.c.}, \quad v_y = 0$$

$$v_z = (-q\varphi_0 e^{-qz} - iqa_0 e^{-\delta z}) e^{-iqx+i\Omega t} + \text{c.c.} \qquad (3.70)$$

From the equation (3.57) taking into account (3.49), (3.68) we have

$$\xi_{20} = -\frac{q\varphi_0 + iqa_0}{i\Omega} \qquad (3.71)$$

We use (3.68), (3.70), (3.71) in (3.66) and seek the stationary value of $T_1(x,z,t)$ in the form

$$T_1(x, z, t) = T_{10}(z) \exp(-iqx + i\Omega t) .$$

The formula for $T_{10}(z)$ is easily found from (3.66), but has a quite combersome form. However, to obtain the dispersion equation, it is sufficient to know $T_1(x, z=0, t)$. Introducing the notations

$$A_1 = \frac{f_1}{\gamma_T\,[\chi(\gamma_T+\gamma_1+\gamma^*)-v_0]} + \frac{f_{-1}}{\gamma_T\,[\chi(\gamma_T+\gamma^*_{-1}+\gamma)-v_0]} -$$

$$- \frac{f_0}{\chi\gamma_T} - \frac{gi\Omega}{\gamma_T\kappa} + \frac{gv_0}{\kappa}\left(1-\frac{v_0}{\chi\gamma_T}\right),$$

$$A_2 = \frac{f_0}{\chi\gamma_T\,[\chi(\gamma_T+\gamma_0+q)-v_0]\,(\gamma_T+q)} + \frac{\chi(\gamma_T+q)-v_0}{\chi(\gamma_T+q)}\,\frac{g}{\gamma_T\kappa},$$

$$A_3 = \frac{f_0}{\chi\gamma_T\,[\chi(\gamma_T+\gamma_0+\delta)-v_0]\,(\gamma_T+\delta)} + \frac{\chi(\gamma_T+\delta)-v_0}{\chi(\gamma_T+\delta)}\,\frac{g}{\gamma_T\kappa},$$

$$\gamma_T = \frac{v_0}{2\chi} + \left(q^2+\frac{i\Omega}{\chi}+\frac{v_0^2}{4\chi^2}\right)^{1/2}, \qquad (T_{10}(z) \sim e^{-\gamma_T z}), \tag{3.72}$$

we write down $T_1(x, z=0, t)$ in the form

$$T_1(x,z=0,t) = \left[\left(A_1-\frac{gv_0}{\kappa}\right)\xi_0 - qA_2\varphi_0 - iqA_3 a_0\right]e^{-iqx+i\Omega t} + \text{c.c.} \tag{3.73}$$

Thus, the temperature field is linear function of initial fluctuation amplitudes ξ_0, φ_0, a_0.

3.4.2. Dispersion equation for relief modulation waves. Competition between three feedback mechanism in the process of surface structures formation in different laser heating regimes

To obtain the temperature distribution on the surface $z = \xi(x, t)$, we shall expand (3.55) in Tailor series up to the terms, proportional to ξ^2. Taking into account the last boundary condition (3.62), we have

$$T(x,z=\xi(x,t), t) = T_{0s} + \frac{gv_0}{\kappa}\,\xi(x,T) + T_1(x,z=0,t).$$

Using this equation, we linearize the relation (3.51) with respect to $\xi(x, t)$. Taking into account (3.65), we have

$$\frac{\partial \xi_1(x,t)}{\partial t} = \left(\frac{g v_0^2 U}{\kappa RT_{0s}^2} - \frac{\sigma v_0}{RT_{0s}} q^2\right) \xi(x,t) + \frac{v_0 U}{RT_{0s}^2} T_1(x,z=0,t) \qquad (3.74)$$

Now we substitute (3.49), (3.68), (3.71), (3.73) into the equation (3.74) and the boundary conditions (3.56). Then we obtain the system of three, linear, homogeneous equations for the initial fluctuating amplitudes ξ_0, φ_0, a_0:

$$(i\Omega + \gamma_1 - \gamma_1 A_1)\xi_0 + (v_T q A_2 + q)\varphi_0 + i(v_T q A_3 + q)a_0 = 0 ,$$

$$\left(\frac{\omega_0^2}{q} - p_{0T} A_1\right)\xi_0 + (p_{0T} q A_2 - i\Omega - 2\nu q^2 - v_0 q)\varphi_0 + i(p_{0T} q A_3 - 2\nu q\delta)a_0 = 0$$

$$-\sigma_{0T} q A_1 \xi_0 + (\sigma_{0T} q^2 A_2 + 2\nu q^2)\varphi_0 + i[\sigma_{0T} q^2 A_3 + \nu(\delta^2 + q^2)]a_0 = 0 , \quad (3.75)$$

where

$$v_T = \frac{v_0 U}{RT_{0s}^2} , \qquad p_{0T} = \frac{p_{in} U}{2\rho RT_{0s}^2} ,$$

$$\gamma_{in} = \frac{v_T \sigma T_{0s} q^2}{U} , \qquad \omega_0^2 = g_0 q + \sigma_0 q^3 + \frac{p_{0T} \sigma T_{0s} q^3}{U} .$$

(ω_0 is the frequency of free CW). Putting the determinant of the system (3.75) equal to zero, we obtain the dispersion equation.

Analysis of the experimental data shows that in all cases of practical interest $\gamma_1 \ll |\Omega|, 4\nu q^2$. Because of this dispersion equation, following from (3.75), has the form

$$(i\Omega)^3 + (i\Omega)^2 [4\nu q^2 + v_0(q+\delta)] + i\Omega [4\nu^2 q^3 (q-\delta) + \omega_0^2 + 2\nu q^2 v_0(q+\delta) +$$

$$+ v_0^2 q\delta] + \omega_0^2 v_0\delta = \left\{ -\sigma_{0T} q^2 [i\Omega + 2\nu q(q-\delta) + v_0 q] + \right.$$

$$+ p_{0T} q [i\Omega + v_0\delta] + v_T [(i\Omega)^2 + i\Omega(4\nu q^2 + v_0(\delta+q)) + 4\nu^2 q^3(q-\delta) +$$

$$\left. + 2\nu q^2 v_0(\delta + q) + v_0^2 \delta q] \right\} A_1 + \left\{ \sigma_{0T} q^2 \left[i\Omega 2\nu q\delta + \omega_0^2 \right] + \right.$$

$$\left. + p_{0T} q \left[(i\Omega)^2 + i\Omega(2\nu q^2 + \delta v_0) \right] - v_T \omega_0^2 \left[i\Omega + \delta v_0 + 2\nu q^2 \right] \right\} A_2 +$$

$$+ \left\{ -\sigma_{0T} q^2 \left[(i\Omega)^2 + i\Omega(2\nu q^2 + v_0 q) + \omega_0^2 \right] - \right.$$

$$\left. - p_{0T} q i\Omega 2\nu q^2 + v_T 2\nu q^2 \omega_0^2 \right\} A_3 \tag{3.76}$$

For pump wave parameters for which $\gamma_q > 0$, the amplitudes of the corresponding Fourier components of the surface relief grow exponentially as a function of time. The rate of this growth is determined by the thermocapillary forces ($\sim \sigma_{0T}$), evaporation pressure forces ($\sim p_{0T}$), and the direct outflow of mass accompanying evaporation ($\sim v_T$). The contributions of these processes to Ω, according to (3.76), are in the approximate ratio $|\sigma_{0T}| q^2 : p_{0T} q : v_T |\Omega|$. For quite low pumping intensities, when the surface temperature $T0s$ is such that the inequality

$$|\sigma_{0T}| q > \frac{p_{in} U}{2\rho R T_{0s}^2} \tag{3.77}$$

holds, the thermocapillary mechanism of excitation of the modulations of the relief dominates. When the condition (3.77) breaks down, which occurs when the pumping intensity is raised, the effect of the pressure of evaporation on the motion of the liquid must be taken into account. As the intensity is further raised, when $|\Omega|$ grows to such an extent that (here we took into account the relation (3.52))

$$|\Omega| > \left(\frac{\pi R T_{0s}}{2\rho} \right)^{1/2} q \tag{3.78}$$

the mass outflow accompanying evaporation is the determining factor.

The complex frequency $\Omega(q)$, according to Eq. (3.76), depends substantially on the parameters of the incident radiation through the quantitites A_1, A_2, A_3. The first two terms in the expression (3.72) for A_1 arise owing to the presence of interference thermal sources (f_1, f_{-1}) in (3.66). Their magnitude depends on the orientation of q relative on the plane of incidence of the laser radiation. The third term in A_1 corresponding in the boundary conditions (3.66) to

102

the term proportional to d^2T_0/dz^2, characterizes the thermal contribution to Ω, which is independent of the orientation of $\mathbf{q}$. The quantities A_2 and A_3 are independent of the orientation of $\mathbf{q}$ and correspond in (3.66) to the term $v_z dT_0/dz$.

3.4.3. Generation of capillar waves and the characteristics of the dominant surface structures for s-polarization of the pumping wave

When the surface temperature is not much greater than the melting temperature, so that the condition (3.78) holds, we can set $v_0 = p_{0T} = v_T = 0$, in (3.69), (3.72), and (3.76), i.e., evaporation effects can be neglected. In this case, after some transformations, from (3.76) we shall obtain the dispersion equation describing the instability of the CW amplitudes with arbitrary polarization of the pump:

$$(i\Omega)^2 + 4i\Omega\nu q^2 + 4\nu^2 q^3(q-\delta) + \omega_0^2 = \frac{a_q(\delta-q)}{\delta+q} -$$

$$-\left[\frac{2q\delta}{\delta^2-q^2} + \frac{\omega_0^2}{(i\Omega)^2}\right] b_q + \left[\frac{\delta^2+q^2}{\delta^2-q^2} + \frac{\omega_0^2}{(i\Omega)^2}\right] c_q , \qquad (3.79)$$

where

$$a_q = |\sigma_{0T}| \left[\frac{f_1 q^2}{\chi\gamma_T(\gamma_T+\gamma_1+\gamma^*)} + \frac{f_{-1} q^2}{\chi\gamma_T(\gamma_T+\gamma^*_{-1}+\gamma)} - \frac{f_0 q^2}{\chi\gamma_T}\right],$$

$$b_q = |\sigma_{0T}| \frac{f_0 q^2(\gamma_T-q)}{\chi\gamma_T(\gamma_T+\gamma_0+q)}$$

$$c_q = |\sigma_{0T}| \frac{f_0 q^2(\delta-\gamma_T)}{\chi\gamma_T(\gamma_T+\gamma_0+\delta)} \frac{\nu}{\chi-\nu} . \qquad (3.80)$$

Here we took into account the fact that $\sigma_{0T} < 0$ by seting $(-\sigma_{0T}) = |\sigma_{0T}|$.
If f_0 and f_1 are given by the formulas (3.60), then Eq. (3.79) describes the rate of growth (decay) of the Fourier amplitudes of the CW with s-polarization of the pump and arbitrary ω.

For simplicity we shall further assume that the conditions

$$\nu \ll \chi, \quad |\epsilon| \gg 1, \quad \gamma_0 \gg |\gamma_T|, \quad |\Omega| \gtrsim \omega_0, \quad |\Omega| \gg \nu q^2 \qquad (3.81)$$

hold. This situation is often realized in experiments (for example, in Refs. [145, 139, and 153]). It is easy to see that under the conditions (3.81) Eq. (3.79) is substantially simplified and assumes the form

$$(i\Omega)^2 = a_q - \omega_0^2 . \qquad (3.82)$$

We substitute (3.60) into (3.80) and use the relation $(k_a^2 - \gamma_a \Gamma_a)(\Gamma_a + \gamma_a) = k_0^2(\epsilon \Gamma_a + \gamma_a)$, $a = \pm 1$. Then the expression for a_q can be written in the form

$$a_q = \frac{|\sigma_{0T}| f_0 q^2}{\chi \gamma_T} \left[\frac{\epsilon - 1}{\gamma_T + \gamma_1 + \gamma^*} \left(\frac{k_1^2 \sin^2 \varphi_1}{\epsilon \Gamma_1 + \gamma_1} - \frac{k_0^2}{\Gamma_1 + \gamma_1} \right) + \right.$$

$$\left. + \frac{\epsilon^* - 1}{\gamma_T + \gamma_{-1}^* + \gamma} \left(\frac{k_{-1}^2 \sin^2 \varphi_{-1}}{\epsilon^* \Gamma_{-1}^* + \gamma_{-1}^*} - \frac{k_0^2}{\Gamma_{-1}^* + \gamma_{-1}^*} \right) - 1 \right] . \qquad (3.83)$$

The number one in the brackets in (3.83) corresponds to the purely thermal, independent of the orientation of q, contribution to Ω. We shall now show that a_q undergoes resonant growth, when $k_1 \approx k_0$ or $k_{-1} \approx k_0$. Keeping in mind the conditions (3.81), we can set in the expression (3.83) near resonance

$$\Gamma_1 + \gamma_1 \approx \gamma_1 \approx (m-in)k_0 , \quad \Gamma_{-1}^* + \gamma_{-1}^* \approx \gamma_{-1}^* \approx (m+in)k_0 ,$$

$$\gamma_T + \gamma_1 + \gamma^* \approx \gamma_T + \gamma_{-1}^* + \gamma \approx 2k_0 m .$$

In this case the sum of the terms proportional to $k_0^2/(\Gamma_a + \gamma_a)$ exactly compensate the unit term in the brackets in (3.83). As a result near resonance we have

$$a_q = \frac{|\sigma_{0T}| f_0 q^2}{\chi \gamma_T 2m} (1_1 \sin^2 \varphi_1 + 1_{-1}^* \sin^2 \varphi_{-1}), \quad 1_a = \frac{\epsilon k_0}{\epsilon \Gamma_a + \gamma_a} .$$

$$a = \pm 1 \qquad (3.84)$$

104

We introduce the notation

$$x_1 = \left(\frac{k_1^2 - k_0^2}{k_0^2}\right)^{1/2} - \beta_m , \qquad y_s = \left(\frac{k_0^2 - k_1^2}{k_0^2}\right)^{1/2} + \beta_n$$

$$\beta_m = \frac{m}{m^2 + n^2} , \qquad \beta_n = \frac{n}{m^2 + n^2} .$$

In this notation the real and imaginary parts of l_1 near resonance ($k_1 \approx k_0$) can be written in the form

$$\mathrm{Re}\, l_1 = \frac{x_1}{x_1^2 + \beta_n^2} , \qquad \mathrm{Im}\, l_1 = \frac{\beta_n}{x_1^2 + \beta_n^2} , \qquad \text{if } k_1 > k_0$$

$$\mathrm{Re}\, l_1 = -\frac{\beta_m}{y_1^2 + \beta_m^2} , \qquad \mathrm{Im}\, l_1 = \frac{y_1}{y_1^2 + \beta_m^2} \qquad \text{if } k_1 < k_0 . \qquad (3.85)$$

The qualitative dependence of $\mathrm{Re}\, l_1$ and $\mathrm{Im}\, l_1$ on k_1 for metals ($m > n$) and dielectrics ($m < n$) are presented in Fig. 28. We shall further assume that $q^2 + (\gamma_q/\chi) > \Omega_q/\chi$, so that ($\gamma_T \approx \gamma_{0T} = [q^2 + (\gamma_q/\chi)]^{1/2}$) may be regarded as a real quantity. Then, substituting (3.85) into (3.84), we obtain in the region $k_a \geqslant k_0$

$$a = \frac{|\sigma_{0T}|\, f_0 q^2}{\chi \gamma_{0T}\, 2m} \left(\frac{x_1 \sin^2 \varphi}{x^2 + \beta_n^2} + \frac{x_{-1}^2 \sin^2 \varphi_{-1}}{x_{-1}^2 + \beta_n^2} \right)$$

$$b = \frac{|\sigma_{0T}|\, f_0\, q^2}{\chi \gamma_{0T}\, 2m} \left(\frac{\beta_n \sin^2 \varphi_1}{x_1^2 + \beta_n^2} - \frac{\beta_n \sin^2 \varphi_{-1}}{x_{-1}^2 + \beta_n^2} \right) , \qquad (3.86)$$

where $a \equiv \mathrm{Re}\, a_q$, $b \equiv \mathrm{Im}\, a_q$. Analogous expressions can be obtained for a and b in the region $k_a < k_0$, in which for metals the quantity b has a local maximum (see Fig. 28 c and d). The expression for f_0 in (3.86) for $|\epsilon| \gg 1$ can be written, according to (3.60), in the form

$$f_0 \approx \frac{4k_0 I_i}{c_v}\, \frac{2mn}{m^2 + n^2}\, \cos^2 \theta , \qquad I_i = \frac{c|E_i|^2}{2\pi} \qquad (3.87)$$

105

In the notation of (3.86) the solution of Eq. (3.82) has be form

$$\gamma_q = \frac{a - \omega_0^2}{2} + \frac{1}{2}\left[(a - \omega_0^2)^2 + b^2\right]^{1/2}$$

$$\Omega_q = \frac{b}{|b|}\left\{-\frac{a - \omega_0^2}{2} + \frac{1}{2}\left[(a - \omega_0^2)^2 + b^2\right]^{1/2}\right\}^{1/2}. \tag{3.88}$$

It is evident from (3.88) and (3.86) that γ_q is maximum for those CW for which a and b^2 assume their maximum values. In the case when $|\epsilon| \gg 1$ (see Fig. 27) a = max, when $k_a^2 = k_0^2 + (m + n)^2 k_0^2/(m^2 + n^2)^2 \approx k_0^2$, and $b^2 = $ max for $k_a^2 = k_0^2 + m^2 k_0^2/(m^2 + n^2) \approx k_0^2$.
This means that for the dominant CW the modulus $|k_a| \approx k_0$ is fixed, and the orientation of k_a for the time being remains arbitrary (see Fig. 27a). In accordance with the law of conservation of momentum (3.1a) the wave vectors of these CW are found from the relation

$$q^2 = k_a^2 + k_t^2 - 2k_a k_t \cos\varphi_a . \tag{3.89}$$

In what follows those CW for which $k_1 = k_{-1}$, $\sin^2\varphi_1 = \sin^2\varphi_{-1}$ will be called degenerate. The degenerate case of the orientation of q (or k_a) is realized for: a) $\theta = 0$ and arbitrary values of φ_1 and b) $\theta \neq 0$ and $\delta_1 = \pi/2$ (see Fig. 27a). All other CW will be called nondegenerate. For degenerate CW two diffracted electromagnetic waves (Stokes and anti-Stokes) make simultaneously the same resonant contribution to a (3.86), and in this case b = 0. In the nondegenerate case only one of the two waves (Stokes and anti-Stokes) can make a resonant contribution to a and b^2.
We introduce the quantity

$$a_1 = \frac{4k_0 \, |\sigma_{0T}| \, I_i q^2 \cos^2\theta \sin^2\varphi_1}{\chi c_v \left[q^2 + (\gamma_q/\chi)\right]^{1/2}} . \tag{3.90}$$

Then, on the basis of the remarks made above, we shall formulate the results as follows, The quantity γ_q for degenerate CW assumes its maximum value when $k_1^2 = k_0^2 + (m+n)^2 k_0^2/(m^2+n^2)^2$ and is equal to (a = max, b = 0)

$$\gamma_q^2 = 0, \quad \Omega_q^2 = \omega_0^2 - a_1 , \qquad \text{if } a_1 \leqslant \omega_0^2$$

$$\gamma_q^2 = a_1 - \omega_0^2, \quad \Omega_q = 0 , \qquad \text{if } a_1 > \omega_0^2 , \tag{3.91}$$

106

where in the expression (3.90) for a_1, according to (3.89), we must set $q \approx k_0 \cos\theta$, $\sin^2\varphi_1 \approx \cos^2\theta$. The maximum values of γ_q for nondegenerate CW with $a_1 \leqslant 24\omega_0^2/7$ are attained when $k_1^2 = k_0^2 + mk_0^2/(m^2 + n^2)^2$ and are equal to $(a = 0, b = \max)$

$$\gamma_q^2 = \frac{1}{2}\left[-\omega_0^2 + (\omega_0^4 + a_1^2)^{1/2}\right], \quad \Omega_q = \frac{a_1}{2\gamma_q} \tag{3.92}$$

and for $a_1 > 24\omega_0^2/7$ the maximum values of γ_q are given by the formula $(k_1^2 = k_0^2 + k_0^2(m + n)^2/(m^2 + n^2)^2, a = \max, b \neq 0)$

$$\gamma_q^2 = \frac{1}{2}\left\{\frac{a_1}{2} - \omega_0^2 + \left[\left(\frac{a_1}{2} - \omega_0^2\right)^2 + \frac{a_1^2}{4}\right]^{1/2}\right\}, \quad \Omega_q = \frac{a_1}{4\gamma_q} \tag{3.93}$$

In the expressions (3.92) and (3.93) a_1 is determined by the formula (3.90). The formulas (3.91) – (3.93) give the frequencies of the dominant CW and the growth rates of the amplitudes as a function of the orientation of q and of the pumping intensity. Figure 30 shows the dependence of γ_q on $\cos\varphi_1$ for three values of θ according to the formulas (3.91) – (3.93.). The absolute maximum of γ_q corresponds to CW with the parameters

$$\mathbf{q} \parallel \mathbf{E}_i, \quad q = q_1 \equiv k_0(n^{*2} - \sin\theta)^{1/2}, \quad n^{*2} = 1 + \frac{(m + n)^2}{(m^2 + n^2)^2} \tag{3.94}$$

It is evident from Fig. 30 that for $\theta > 35$–40^0 γ_q assumes its absolute maximum value when $\cos\varphi_1 \approx 0$, i.e., for the nondegenerate wave with the parameters (see Fig. 27)

$$q \approx k_0(1 + \sin^2\theta)^{1/2}, \quad \cos\delta_1 = -\frac{\sin\theta}{(1 + \sin^2\theta)^{1/2}} \tag{3.95}$$

In this case the vector $\mathbf{q}$ of the wave (3.95) is neither parallel nor perpendicular to $\mathbf{E}_i$. We emphasize that Fig. 30 is more qualiative than quantitative in nature, since in calculating γ_q we set $q^2 + (\gamma_q/\chi) \approx q^2$ in (3.90).

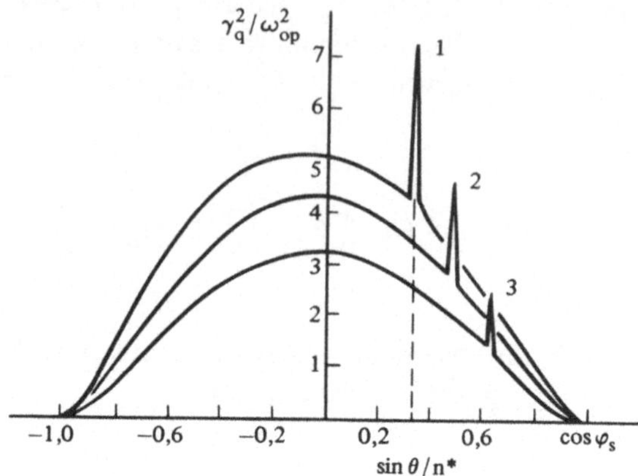

Fig. 30. Theoretical dependence of the growth rate γ_q of the dominant CW in a germanium melt on the orientation of q and the angle of incidence of the s-polarized pump wave according to the formulas (3.91)–(3.93). T_{melt} = 1200 K, ρ = 5.51 g/cm^3, ν = 1.35·10^{-2} cm^2/s. χ = 0.1 cm^2/s, c_v = 2.45·10^7 erg/cm^3·deg [181]; σ = 600 dynes/cm, $\sigma_0 T$ = 0.07 cm^3/s^2·deg [182]; ϵ = −32 + i 72 [153]. Pumping parameters λ = 1 μm. I_1 = 2·10^7 W/cm^2, ω_{op} = 1.64·10^3 c^{-1}, 1 − θ = 20°; 2 − θ = 30°, 3 − θ = 40°.

We note that for both metals and dielectrics the growth rates of the dominant CW (and this is evident from (3.89)–(3.93)) are virtually independent of n and m. This occurs because of the fact that the dependence of the quantities a and b on m and n, appearing in (3.86) through f_0, is compensated by the dependence of the resonant factor (3.85) on m and n.

3.4.4. Generation of capillar waves and the characteristics of the dominant surface structures for p-polarization of the pumping wave

In this case the quantities f_0 and f_1 in Eqs. (3.79) and (3.80) are given by the formulas (3.61). We shall study here metals (m $>$ n) and dielectrics (m $<$ n) for which $|\epsilon| \gg 1$. As in the case of s-polarization of the pump wave the quantity f_{-1} (3.61) exhibits a resonance near $k_{-1} \approx k_0$. Near the resonance the expression (3.61) for f_1 can be written in the form

$$f_1 = f_0(\epsilon-1)\frac{k_0^2\, f(\theta,\varphi_1)}{\epsilon\Gamma_1+\gamma_1} - \frac{k_0^2}{\Gamma_1+\gamma_1} \tag{3.96}$$

108

where $f(\theta, \varphi_1) = n^* \cos^2\varphi_1 - \sin\theta \cos\varphi_1$, $n^* = k_1/k_0 \approx 1$, while the quantity f_0, unlike the corresponding expression (3.87), right up to angles of incidence θ close to the glancing angle, is independent of θ and is equal to

$$f_0 = \frac{4k_0 I_i}{c_v}\, \frac{2mn}{m^2 + n^2}\,.$$
(3.97)

The expression for f_{-1} is obtained from (3.96) with the help of the substitution indicated after the formula (3.61).

We shall assume that the conditions (3.81) hold. In this case the dispersion equation (3.79) reduces to the form (3.82), where

$$a_q = \frac{|\sigma_{0T}| f_0 q^2}{\chi \gamma_T}\left[\frac{\epsilon - 1}{\gamma_T + \gamma_1 + \gamma^*}\left(\frac{k_0^2 f(\theta, \varphi_1)}{\epsilon \Gamma_1 + \gamma_1} - \frac{k_0^2}{\Gamma_1 + \gamma_1}\right) + \right.$$

$$\left. + \frac{\epsilon^* - 1}{\gamma_T + \gamma_{-1}^* + \gamma}\left(\frac{k_0^2 f(\theta, \varphi_{-1})}{\epsilon^* \Gamma_{-1}^* + \gamma^*} - \frac{k_0^2}{\Gamma_{-1}^* + \gamma_{-1}^*}\right) - 1\right].$$
(3.98)

Comparison of the formulas (3.83) and (3.98) shows that the expression (3.98) is obtained from (3.83) by the substitution f_0 (3.86) $\rightarrow$ f_0 (3.93) and $\sin^2\varphi_{-1} \rightarrow f(\theta, \varphi_a)$. This means that the expressions for the growth rate of the dominant CW in the case of a p-polarized pumping wave will be given by the formulas (3.91)–(3.93), in which (3.90) must be replaced by

$$a_1 = \frac{4k_0 |\sigma_{0T}| I_i\, q^2}{\chi c_v\, [q^2 + (\gamma_q/\chi)]^{1/2}}\, |n^* \cos^2\varphi_s - \sin\theta \cos\varphi_s|.$$
(3.99)

We note that when $\theta \neq 0$, in the situation shown in Fig 27a $f(\theta, \varphi_a) = 0$.

Graphs of the dependence of the growth rate of the dominant CW on the orientation of their wave vectors q (i.e., on $\cos\varphi_1$), shown in Fig. 31, were constructed based on the formulas (3.89), (3.91)–(3.93), and (3.95). In addition, we set in (3.99) $q^2 + \gamma_q/\chi \approx q^2$. It is evident from these graphs that for small fixed angles θ the CW with $\cos\varphi_1 = \pm 1$ have the largest values of γ_q. The parameters of these CW can be described by the formulas

$$\cos\varphi_1 = 1, \quad q \parallel E_{it}, \quad q = k_0\,(n^* - \sin\theta) \equiv q_1$$

$$\cos\varphi_1 = -1, \quad q \parallel E_{it}, \quad q = k_0\,(n^* + \sin\theta) \equiv q_2\,,$$
(3.100)

where n^{*2} in accordance with (3.58) and (3.59) is equal to $n^{*2} = 1 + m^2/(m^2 + n^2)^2$ for $a_1 > 24\omega_0^2/7$ and $n^{*2} = 1 + (m+n)^2/(m^2 + n^2)^2$ for $a_1 > 24\omega_0^2/7$. As the angle θ is increased the growth rate of CW with q_2 increases, and that with q_1 decreases. The growth rates of structures with the parameters $0 < \cos\varphi_1 < n^{*-1} \sin\theta$ increase with the angle θ.

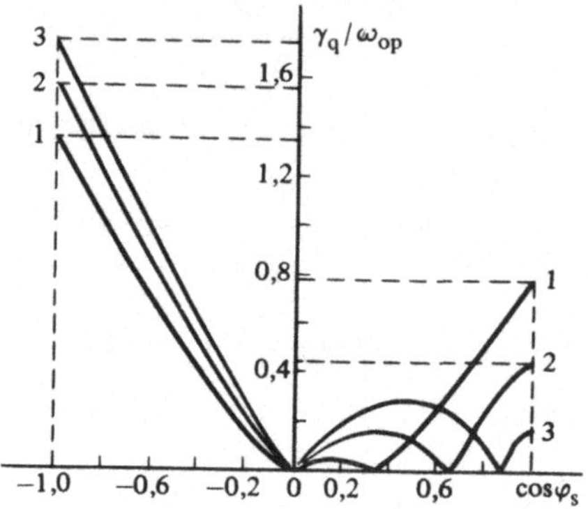

Fig. 31. Dependence of the growth rate of CW γ_q on the orientation of **q** with p polarization of the pump wave for the case of germanium. The parameters are the same as in Fig. 30 (according to the formulas (3.89), (3.91)–(3.93), (3.95)). 1) $\theta = 20^{\circ}$; 2) $\theta = 40^{\circ}$; 3) $\theta = 60^{\circ}$.

3.4.5. Nature of doublet structures

Experiments [178, 179, 153] have established that together with the structures (3.94) and (3.100) structures with the same orientation but with somewhat smaller values of q are generated, i.e., a doublet of structures with close values of q is excited. We shall show (see Ref. [187]) that this theory predicts generation of doublets of CW, whose wave vectors coincide with the experimental values of q for doublets of surface gratings [153,3]. We shall do this for the case of s-polarization of the pump wave and $\delta_s = \pi/2$ (see Fig. 27a) — conditions corresponding to the experiment in Ref. [153].

We shall assume that the conditions (3.81) hold; in this case Eq. (3.79) has the form (3.82), where a_q is given by the formula (3.33). We shall take into account the fact that for $\delta_1 = \pi/2$, $k_1 = k_{-1}$, $\sin^2\varphi_1 = \sin^2\varphi_{-1}$. Then arguments analogous to those in Sec. 4.3 enable writing a_q near resonance ($k_1 > k_0$) in the form

$$a_q \approx \frac{|\sigma_{0T}| f_0 \, q^2}{\chi \gamma_T m} \frac{x_1 \sin^2\varphi_1}{(x_1^2 + \beta_n^2)},$$

110

$\sin^2\varphi_1 \approx \cos^2\theta$. Let $q^2 \gg |\Omega|/\chi$; then $\gamma_T^{-1} \approx q^{-1} - (i\Omega/2\chi q^3)$ and from (3.82) we obtain $(i\Omega)^2 + 2A_q i\Omega + \omega_0^2 - a_q^{(0)} = 0$; $a_q^{(0)} = |\sigma_{0T}| f_0 k_0 x_s \cos^3\theta/\chi\, m(x_1^2 + \beta_n^2)$, $A_q = |\sigma_{0T}| f_0 x_s \cos\theta/4\chi^2 k_0 m(x_1^2 + \beta_n^2)$, where we have set $q \approx k_0\cos\theta$. The solution of this equation has the form

$$\gamma_q + i\Omega_q = -A_q + (A_q^2 + a_q^{(0)} - \omega_0^2)^{1/2} \tag{3.101}$$

We note that (3.101) is a generalization of the formulas (3.91), in whose derivation it was assumed for simplicity that γ_T is real.

For high pumping intensities, such that the condition $\max\left\{A_q^2 + a_q^{(0)}\right\} > \omega_0^2$ holds, the growth rate γ_q has two positive peaks. The first peak in γ_q, determined by the minimum of the quantity Rel_1 (see Figs. 28a and b), corresponds to CW with the parameters

$$q \equiv k_0\,(n^{*2} - \sin^2\theta)^{1/2}, \qquad q \parallel E_i, \tag{3.102}$$

where in the case of metals $n^{*2} = 1 + (m-n)^2/(m^2 + n^2)^2$, while for dielectrics $n^{*2} = 1$. In this case $\gamma_q = -A_q$, $\Omega_q = (\omega_0^2 - a_q^{(0)} - A_q^2)^{1/2}$. The second peak in γ_q is associated with the maximum of Rel_1 (see Figs. 28a and b). The frequency of the corresponding CW is $\Omega_q \approx 0$, and the vector q is determined by the formulas (3.96). The numerical solution for the full equation (3.79) for the case of germanium with $\delta_1 = \pi/2$ is given in Figs. 32 and 33. It is evident

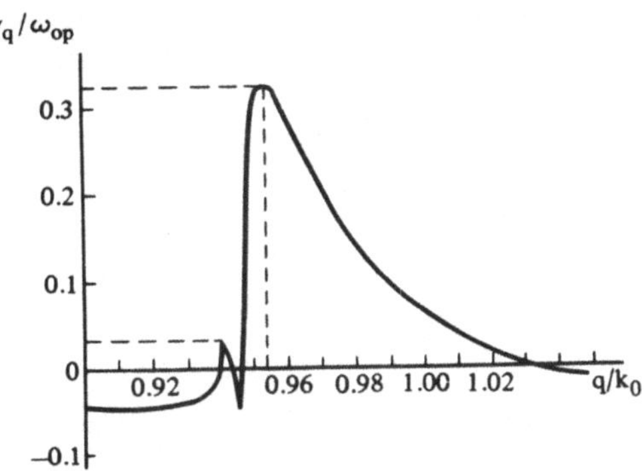

Fig. 32. Dependence of the growth rate of CW γ_q on the modulus of q for $\delta_s = \pi/2$ and an s-polarized pump in the case of germanium. The parameters of the material and of the pump are the same as in Fig. 30 (the numerical solution of Eq. (3.79)).

111

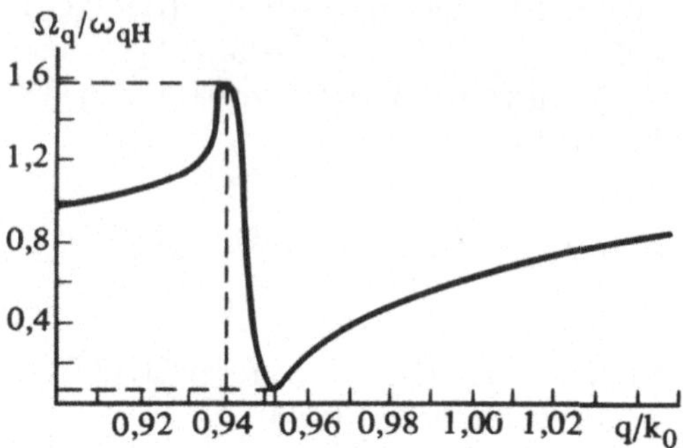

Fig. 33. The dependence of the frequency of CW Ω_q on the modulus of q. The parameters are the same as in Fig. 30 (numerical solution of Eq. (3.79)).

that the peak values of the growth rate, obtained by the numerical calculation (see Fig. 32) and from the approximate formulas (3.81)–(3.83) (see Fig. 30), are somewhat different. For high pumping intensities, when the conditions (3.81) are better satisfied, these values converge to one another. It is evident from Fig. 32 that the magnitude of the peak and the width of the resonance for CW with large values of q is greater than the corresponding values of γ_q for CW with smaller values of q. The positions of the peaks are well described by the expressions (3.102) and (3.94). The experimental [153] and theoretical values of q (according to the formulas (3.94) and (3.102) as a function of θ are compared in Fig. 34. It is evident that they are in good agreement with

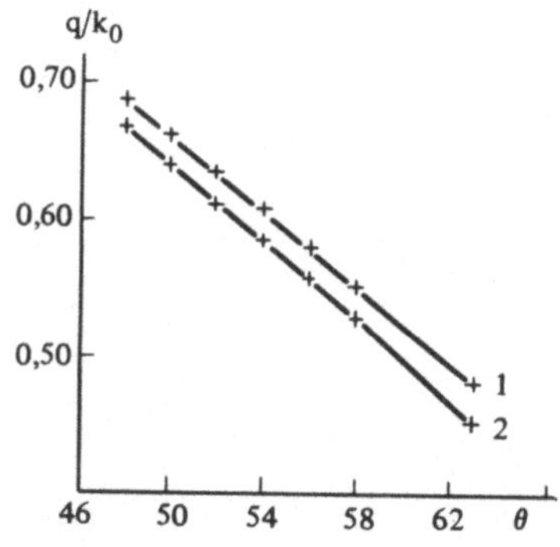

Fig. 34. Comparison of the theoretical (according to formulas (3.74) and (3.82) – solid lines) and experimental [153] (+) dependences of the doublet of values of q on θ in the case of germanium. The parameters are the same as in Fig. 30.

one another. The appearance of doublets of gratings can also be explained by
other mechanisms, which take into account the presence of two extrema in
the electromagnetic factor (see Fig. 28). Thus it is concluded in Refs. [153]
and [202] based on the experiments performed that generation in the upper
branch of Fig. 34 is determined by CW and generation in the lower branch is
determined by the appearance of a nonuniform melt on the surface (see also
Ref. [209]). In Ref. [127] the formation of doublet structures is interpreted
differently — the structures are interpreted as being the consequence of splitting
of the dispersion curve of SEW.

3.4.6. Generation of small-scale structures under the action of laser radiation [187]

Thus far we have studied generation of CW under the condition that $q \sim k_0$.
We shall show that Eq. (3.79) also describes excitation of CW with $q \gg k_0$
(small-scale structures, observed experimentally [146, 175].
we shall assume that the conditions

$$\nu \ll \chi, \quad \nu q^2 \ll |\Omega| \ll \chi q^2, \quad |\Omega|^2 \gtrsim \omega_0^2 \gg \nu q^2 \tag{3.103}$$

hold. In this case the general equation (3.79) simplifies and has the form (3.82),
where a_q in the case of s-polarization (to whose analysis we shall confine our
attention here) is given by the formula (3.83). In the region $q \gg k_0$ in (3.83)
we can set $\Gamma_1 \approx \Gamma_{-1} \approx k_a \approx q$, $\gamma_1 \approx \gamma_{-1} \approx (q^2 - k_0^2 \epsilon)^{1/2}$, $\cos^2 \varphi_1 \approx \cos^2 \varphi_{-1}$.
In addition, in accordance with (3.103) in the formula (3.83) we substitute
$\gamma_T \cong (1/q) - (i\Omega/2\chi q^3)$. Then after some transformations Eq. (3.82) acquires
the form

$$(i\Omega)^2 + A_q \, i\Omega - (a_q^{(0)} - \omega_0^2) = 0 ,$$

$$a_q^{(0)} = \frac{|\sigma_{0T}| f_0 \, q}{\chi} \left\{ 2 \, \mathrm{Re}\left[\frac{\epsilon - 1}{q + \gamma_1 + \gamma^*} \, \frac{\gamma_1 \Gamma_1 \sin^2 \varphi_1}{\epsilon \Gamma_1 + \gamma_1} - \frac{k_0^2 \cos^2 \varphi_1}{\Gamma_1 + \gamma_1} \right] - 1 \right\} ,$$

$$b_q^{(0)} = \frac{|\sigma_{0T}| f_0 \, q}{\chi} \, 2 \, \mathrm{Re}\left[\frac{q(\epsilon - 1)}{(q + \gamma_1 + \gamma^*)^2} \, \frac{\gamma_1 \Gamma_1 \sin^2 \varphi_1}{\epsilon \Gamma_1 + \gamma_1} - \frac{k_0^2 \cos^2 \varphi_1}{\Gamma_1 + \gamma_1} \right] ,$$

$$A_q = \frac{a_q^{(0)} + b_q^{(0)}}{2\chi q^2} . \tag{3.104}$$

where f_0 is given by the formula (3.87). We shall now study separately the following cases:

$$k_0 \ll q \ll k_0 m , \qquad m \gg 1 , \tag{3.105}$$

$$q \gg k_0 |\epsilon|^{1/2} , \qquad |\epsilon| \gg 1 \tag{3.106}$$

In the region determined by the inequalities (3.105), from (3.104) we obtain $(\omega_0^2 - a_q^{(0)} \gg Q_q^2/4)$

$$\Omega_q \approx (\omega_0^2 - a_q^{(0)})^{1/2} , \qquad \gamma_q \approx \frac{|\sigma_0 T| f_0}{4 \chi^2 m k_0} \cos^2 \varphi_1 ,$$

$$a_q^{(0)} \approx - \frac{|\sigma_0 T| f_0 q^2}{\chi m k_0} \left(\frac{3}{2} - \sin^2 \varphi_1 \right) . \tag{3.107}$$

In the region determined by the formulas (3.106), from Eq. (3.104) we have

$$\Omega_q \approx (\omega_0^2 - a_q^{(0)})^{1/2} , \qquad \gamma_q \approx \frac{|\sigma_0 T| f_0}{4 \chi^2 q} (\cos^2 \varphi_1 - \frac{1}{2} \sin^2 \varphi_1) ,$$

$$a_q^{(0)} = - \frac{|\sigma_0 T| f_0 q}{\chi} \cos^2 \varphi_1 . \tag{3.108}$$

Analysis of the formulas (3.107) and (3.108) shows that for arbitrary angles of incidence θ in the region $q \gg k_0$ the CW for which $\cos^2 \varphi_1 = 1$ should have the highest values of γ_q. These CW, unlike the previously studied structures, correspond to waves with orientation $q \perp E_i$. Comparison of (3.107) and (3.108) shows that the transition from (3.107) to (3.108) occurs at $q \approx k_0 m$. In this case, in the region $k_0 \ll q \ll k_0 m$ the quantity γ_q is independent of q, and in the region $q > k_0 m$ it decreases as q increases. Therefore, the continuum of dominating structures should have the parameters

$$q \perp E_i , \qquad k_0 \leqslant q < k_0 m . \tag{3.109}$$

In the case of a germanium melt with $\lambda = 1.06$ μm, $\epsilon = -32 + i72$ [153]. Using this expression for ϵ, in accordance with Drude's model we find that for $\lambda = 10.6$ μm $\epsilon = -38.9 + i870$, i.e., $n = 20.4$ and $m = 21.3$. Therefore at $\lambda = 10.6$ μm the periods of the frozen dominant "fine" structures must lie, according to (3.109), in the region $\lambda/m \cong 0.5$ μm $\leqslant d \leqslant \lambda = 10.6$ μm. This is in good agreement with experiment [146], where independently of θ a continuum of structures with $0.8~\mu\text{m} < d < 3~\mu\text{m}$ and the orientation (3.109) was generated.

We shall calculate γ_q from the formula (3.107) with the characteristic experimental intensity $I_i = 10^8~\text{W/cm}^2$ ($\lambda = 10.6~\mu\text{m}$, $\tau_p = 2\cdot 10^{-7}\text{s}$ [146]). According to (3.87) and (3.107), we have at $\theta = 0$:

$$\gamma_q = \frac{|\sigma_{0T}|\,I_i}{\chi^2\,c_v}\,\frac{2n}{m^2+n^2}\,.$$

For estimates we shall use the data which are presented in the caption to Fig. 30, and the value of ϵ calculated based on Drude's model. For these parameters we obtain $\gamma_q = 1.35\cdot 10^7~\text{s}^{-1}$, i.e., $\gamma_q\tau_p = 2.7$. This means that the mechanism of the instability of CW gives rise to substantial growth of the amplitudes of small scale CW over the time τ_p. For p-polarization the results are analogous.

3.5. Dependence of the orientations and periods of the surface ripples generated due to the interference instability on the dielectric permeability of the medium

At sufficiently high pump intensitis (when the condition (3.78) holds) the evaporation waves begin to play the main role in the SPS formation. In this case the relief modulation wave is given as before by the formula (3.49), where now $\xi_1(t) \gg \xi_2(t)$. The dispersion equation for evaporation wave is derived most simply from the system (3.75), neglecting in it the hydrodynamic terms ($\varphi_0 = a_0 = 0$). Then from the first equation of (3.75) we have for the complex frequency (using the condition $v_T A_1 \gg \gamma_1$).

$$i\Omega = v_T\,A_1 \tag{3.110}$$

This dispersion equation may be also derived from the general dispersion equation (3.76), when $|\Omega| \gg vq^2$, $v_0 q$, ω_0. The gratings with maximum value of γ_q will be dominant in the process of relief generation (in liear regime). We will show in this section that the periods and orientations of such gratings depend essentially on the value of dielectric permeability of the medium $\epsilon(\omega)$.

3.5.1. s-polarization of the pump wave

Substituting v_T and A_1 from (3.75) and (3.72) into Eq. (3.110) and using the relation $(\Gamma_a + \gamma_a)(k_a^2 - \gamma_a\Gamma_a) = k_0^2(\epsilon\Gamma_a + \gamma_a)$, we obtain an expression for the complex increment $i\Omega$, which is valid for arbitrary values of $\epsilon(\omega)$ and arbitrary angles of incidence θ. Under the conditions that $q^2 \sim k_0^2 \gg v_0^2/\chi^2$, $|\Omega|/\chi$, the expression for the complex increment has the form:

$$i\Omega = \frac{2\omega\epsilon''}{\pi g} \frac{k_z^2 |E_{ix}|^2}{|k_z + i\gamma|^2} \left[\frac{\epsilon - 1}{\gamma_1 + \gamma^* + q} \left(\frac{\gamma_1 \Gamma_1 \sin^2\varphi_1}{\epsilon\Gamma_1 + \gamma_1} - \frac{k_0^2 \cos^2\varphi_1}{\Gamma_1 + \gamma_1} \right) + \right.$$

$$\left. + \frac{\epsilon^* - 1}{\gamma_{-1}^* + \gamma + q} \left(\frac{\gamma_{-1}^* \Gamma_{-1}^* \sin^2\varphi_{-1}}{\epsilon^*\Gamma_{-1}^* + \gamma_{-1}^*} - \frac{k_0^2 \cos^2\varphi_{-1}}{\Gamma_{-1}^* + \gamma_{-1}} \right) \right] . \tag{3.111}$$

The equation in the form (3.111) is valid, if $g/\kappa \gg \dfrac{RT_{0s}^2\delta}{Uv_0}$. If on the contrary, $g/\kappa \ll RT_{0s}^2\delta/U\delta$, than in (3.111) one must replace $\dfrac{g}{\kappa} \rightarrow \dfrac{RT_{0s}^2\,\delta}{Uv_0})$.

The equation (3.111) contains two types of terms, proportional to $\sin^2\varphi_a$ and $\cos^2\varphi_a$ reflecting the peculiarities of diffraction on the surface ripples.
In the case of normal incidence ($\theta = 0$) in (3.111) $\Gamma_1 = \Gamma_{-1}$, $\gamma_1 = \gamma_{-1}$, $k_1 = k_{-1} = q$, $\cos^2\varphi_1 = \cos^2\varphi_{-1}$. Then the growth rate γ_q can be represented in the following general form

$$\gamma_q = \mathrm{Re}(i\Omega) = \frac{2\omega\epsilon''}{\pi g} \frac{k_0^2 |E_{ix}|^2}{|k_0 + i\gamma|^2} 2\,\mathrm{Re}\left[\frac{\epsilon - 1}{\gamma_1 + \gamma^* + q} \left(\frac{\gamma_1 \Gamma_1 \sin^2\varphi_1}{\epsilon\Gamma_1 + \gamma_1} - \right.\right.$$

$$\left.\left. - \frac{k_0^2 \cos^2\varphi_1}{\Gamma_1 + \gamma_1} \right) \right] = b\,(q) \sin^2\varphi_1 + a\,(q) \cos^2\varphi_1 . \tag{3.112}$$

Let us divide the whole range of values of ϵ into four seperate rate regions: a) $\epsilon' < -1$, b) $0 < \epsilon' < 1$, c) $-1 \leq \epsilon' < 0$, d) $\epsilon' > 1$ and study the equation (3.111) in each of these regions seperately.
a) Surface polariton region $\epsilon' < -1$, ($|\epsilon'| \gg \epsilon''$).
One may easily verify, that the main contribution to (3.111, 112) give the terms proportional to $\sin^2\varphi_a$ which describe the exitation of SEW (at resonance $b/a \gg 1$). Retaining only these terms, and using for resonance factor in (3.111) the equation (3.12), one obtains

$$\gamma_q = \frac{\omega\epsilon''k_0 |E_{ix}|^2\cos^2\theta}{\pi g |\epsilon'|^2} \mathrm{Re}\left[\frac{\Gamma_p \sin^2\varphi_1}{\Delta k_1 - i\Gamma_p} + \frac{\Gamma_p \sin^2\varphi_{-1}}{\Delta k_{-1} + i\Gamma_p} \right] . \tag{3.113}$$

116

One can see from this equation, that the grating, for which the resonance condition $\Delta k_a = \Gamma_p$, i.e.

$$k_a = k_0 \sqrt{\frac{|\epsilon'|}{|\epsilon'| - 1}} \left(1 + \frac{\epsilon''}{2|\epsilon'|^2}\right) \approx k_0 \qquad (3.114)$$

is fulfilled, will grow most rapidly.

Thus in the momentum conservation equation (3.1a) the modulus of vector k_a is fixed, but its direction remain unspecified. This means that the ends of all possible vectors k_a lie on the circle with radius $\sim k_0$ (Fig. 27). The conservation law (3.1) with given k_a determines vector q (Fig. 27).

$$q = k_0 \sqrt{1 + \sin^2\theta - 2\sin\theta \cos\varphi_a}, \quad \cos\delta = \frac{k_0}{q}(\cos\varphi_a - \sin\theta) \quad (3.115)$$

All possible cases of mutual orientations of vectors k_a and k_t can be devided into two classes. **Two-resonance case**: the resonance condition (3.114) is fulfilled for both vectors k_1 and k_{-1}. This means, that both terms, in (3.113), corresponding to k_1 and k_{-1} give the same resonance contribution into (3.113). When $\theta \neq 0$, the two-resonance case is possible only for $q \perp k_t$, i.e. at $\cos\varphi_1 = \sin\theta$ (Fig. 27). In this case from (3.113) one then obtains for grating with $\cos\varphi_1 = \sin\theta$ the following growth rate

$$\gamma_{q1} = \frac{|E_{ix}|^2 \omega \cos^2\theta}{\pi g} 2\cos^2\theta \qquad (3.116)$$

At normal incidence ($\theta = 0$) the two-resonance case is realized at any orientation of k_1 (at any value of $\cos\varphi_1$). In this case from (3.113) one has

$$\gamma_{q2} = \frac{|E_{ix}|^2 \omega}{\pi g} 2\sin^2\varphi_1 \qquad (3.117)$$

One-resonance case: $\theta = 0$, the angle between q and k_t is not equal to 90°, and the resonance condition (3.114) is fulfilled only for one vector: either k_1 or k_{-1} (see Fig. 27). In the case of k_1-resonance one can neglect in (3.113) the $a = -1$ contribution and obtains ($\Delta k_1 = \Gamma_p$)

$$\gamma_{q3} = \frac{|E_{ix}|^2 \omega \cos^2\theta}{\pi g} \sin^2\varphi_1 = a_1 \sin^2\varphi_1 \qquad (3.118)$$

The growth rate dependence (3.116), (3.118) on $\cos\varphi_1$ at $\theta \neq 0$ is shown on Fig. 35a. The analisis of obtained results enables to draw the following conclusions:

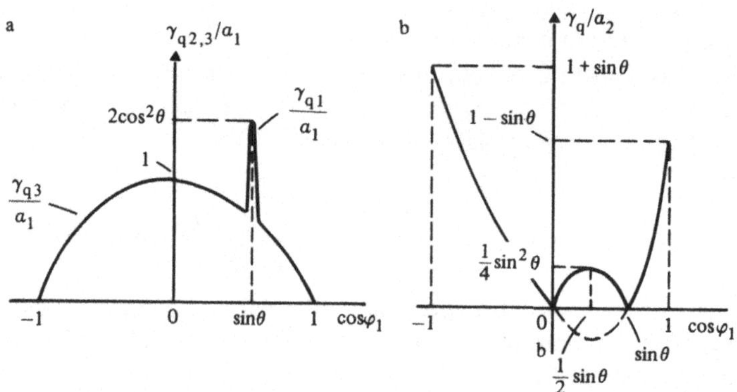

Fig. 35. The growth rate of IEI ($\epsilon' < -1$) dependence on φ_1 a) s-polarization, $\gamma_{q1,2}$ – (3.116), (3.118). b) p-polarization, γ_q – (3.135).

1) The continuum of gratings with the wave vectors $\mathbf{q}$, determined by the equation (3.115), can be generated (i.e. $\gamma_q > 0$) in the case of s-polarized radiation incident at angle θ.

2) If $\theta \neq 0$ (see Fig. 35a) and $2\cos^2\theta > 1$, ($\theta < 45^o$), the absolute maximum of the growth rate is reached at $\cos\varphi_1 = \sin\theta$, this means that the grating with the growth rate (3.116) will be generated most effectively, the period and direction of which, according to (3.115), will be determined by the formula

$$\mathbf{q} \perp \mathbf{k}_t, \qquad d = 2\pi/q = \lambda/\cos\theta \qquad (3.119)$$

3) If $\theta > 45^o$, then the absolute maximum of growth rate is reached at $\cos\varphi_1 = 0$ and two gratings with vectors $\mathbf{q}$, for which $\mathbf{k}_1 \perp \mathbf{k}_t$ will be generated most effectively. The directions and periods of such gratings, according to (3.115) are determined by following expressions

$$d = \lambda/\sqrt{1+\sin^2\theta}, \qquad \cos\delta = -\sin\theta/\sqrt{1+\sin^2\theta} \qquad (3.120)$$

The superposition of two gratings, corresponding to (3.120), may lead to formation of two-dimensional surface structure. Further, due to the broad maximum at $\cos\varphi_1 = 0$ (see (3.118)), the continuum of gratings with parameters near to (3.120) may be generated. In region of values of angles $\theta \approx 45^{\circ}$, when the both maxima Fig. 35a have almost the same height, the superposition of gratings, determined by (3.119) and (3.120), must be observed.

4) At $\theta = 0$, the continuum of gratings with growth rate (3.117) will be generated. The gratings with parameters near to that of grating with $\cos\varphi_1 = 0$ will be most intensive. The period of such grating $d = \lambda$, and wave vector $\mathbf{q} \parallel \mathbf{E}_{ix}$.

5) The value of growth rate deminishes with growth of θ. Let us note that the above considered two-resonance and one-resonance case classification, has been made, in fact, supposing $\Gamma_p \to 0$. In the case of finite values of Γ_p, the grating must be considered to be two-resonance, if

$$\sin\theta - a\Gamma_p/4k_0 \sin\theta \leqslant \cos\varphi_s \leqslant \sin\theta + a\Gamma_p/4k_0 \sin\theta \, , \qquad (3.121)$$

where a is a constant of order unity. Then the smooth transition between the values of growth rate (3.116)–(3.118) occurs. The condition (3.121) was derived by the replacement of resonance condition (3.114) by more mild resonance condition $\Delta k_1 - \Gamma_p \leqslant a\Gamma_p$. The comparison of Fig. 29a, Fig. 30a and Fig. 35a shows that for s-polarization the salient features of growth rate dependences on $\cos\varphi_1$ in the cases of SAW or CW generation and IEI are the same, that is all three intabilities in the linear regime lead to generation of practically the same dominant gratings. The same conclusion is valid and in the case of p-polarization (see below). For the convenience of the reference we shall call the grating with $\mathbf{q} \parallel \mathbf{E}_t$ the longitudinal grating and that with $\mathbf{q} \perp \mathbf{E}_t$ – the transverse one. We thus may conclude that at $\epsilon < -1$ in the case of s-polarization all three istabilities lead to generation of dominant longitudinal gratings.

b) $0 < \epsilon' < 1, \epsilon' \gg \epsilon''$. For the sake of simplicity we will consider explicitly normal incidence case $\theta = 0$, and then generalize the results to the $\theta \neq 0$.

One can easily show, that at each fixed value q from the interval $0 < q < \infty$ in (3.112) the inequality $a(q, \epsilon) > b(q, \epsilon)$ takes place. This means that at any fixed value of q the value of growth rate γ_q reaches maximum, when $\cos^2\varphi_1 = 1$, i.e. the transverse gratings $(\mathbf{q} \perp \mathbf{E}_{ix})$ will grow most rapidly. For transverse gratings for which $\cos^2\varphi_1 = 1$ from (3.112) we obtain

$$\gamma_q = \frac{2\omega\epsilon''}{\pi g} \frac{1 - \sqrt{\epsilon'}}{1 + \sqrt{\epsilon'}} |E_{ix}|^2 \, y(q) \, , \qquad (3.122)$$

$$
y(q) = \begin{cases} k_0^2/q(\Gamma_1 + \gamma_1), & \text{if} \quad q > k_0, \\[2em] \dfrac{\gamma_1(\gamma_1 + q) + \Gamma_{10}k_0\sqrt{\epsilon'}}{(1-\epsilon')q(\gamma_1 + q)}, & \text{if} \quad k_0\sqrt{\epsilon'} \leqslant q \leqslant k_0, \\[2em] k_0/(\Gamma_{10} + \gamma_{10})\sqrt{\epsilon'}, & \text{if} \quad 0 \leqslant q \leqslant k_0\sqrt{\epsilon'}, \end{cases}
$$

where $\Gamma_1 = \sqrt{q^2 - k_0^2}$, $\gamma_1 = \sqrt{q^2 - k_0^2\epsilon'}$, $\Gamma_{10} = \sqrt{k_0^2 - q^2}$, $\gamma_{10} = \sqrt{k_0^2\epsilon' - q^2}$. The analysis of (3.122) shows, that γ_q reaches maximum at $q = k_0\sqrt{\epsilon'}$, if $0 < \epsilon' < 0{,}5$ and at $q \approx 1{,}04\, k_0\sqrt{\epsilon'}$, if $0{,}5 < \epsilon' < 1$. The dependence $\gamma(q)$ is represented on Fig. 36 for two values of $|\epsilon'|$: $|\epsilon'| = 0{,}25$ and $|\epsilon'| = 0{,}81$. Thus, at normal incidence, in the case,

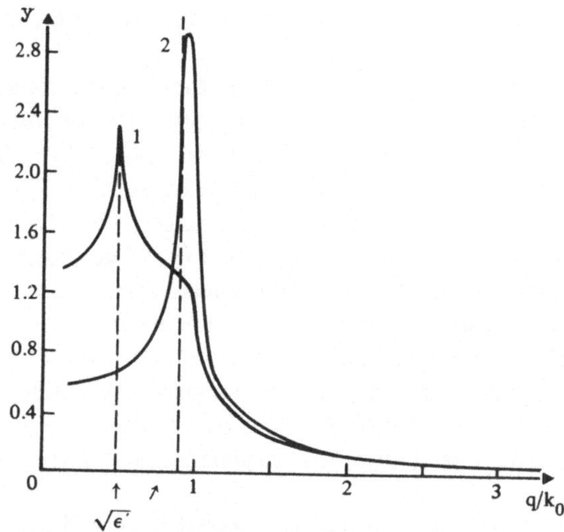

Fig. 36. The characteristic dependences of the growth rate of the transverse gratings on q for $\theta = 0$ in the region $0 < \epsilon' < 1$. (According to the formula (3.122)). 1) $|\epsilon'| = 0.25$; 2) $|\epsilon'| = 0.81$.

when $0 < \epsilon' < 1$, the gratings for which

$$\cos^2\varphi_1 = 1, \quad \text{i.e.} \quad \mathbf{q} \perp \mathbf{E}_{ix}, \quad \mathbf{q} \parallel \mathbf{k}_t \tag{3.123}$$

$$\gamma_1 \approx 0, \quad \text{i.e.} \quad q \approx k_0\sqrt{\epsilon'}, \tag{3.124}$$

120

will grow in time most rapidly. In this case the value of Γ_1, according to (3.6) is imaginary. Thus the diffracted wave (see (3.5)), participating in the generation of the gratings (3.123), (3.124), bears the bulk character in vacuum ($z \leqslant 0$) and is weakly decaying in medium ($z \geqslant 0$) (taking into account $\epsilon'' \neq 0$).

In the case $\theta \neq 0$ from the general expression (3.111) it follows that two transverse gratings with parameters: $q \approx k_0 |\sqrt{\epsilon'} \pm \sin\theta|$, $q \perp E_{ix}$ must be generated.

c) $-1 < \epsilon' < 0$, $|\epsilon'| \gg \epsilon''$. The analysis shows, that at every fixed q in (3.112) the inequality $a(q, \epsilon) > b(q, \epsilon)$ is fulfilled, i.e. the maximum of γ_q is again reached at transverse grating orientation:

$$\cos^2 \varphi_1 = 1, \quad \text{i.e.} \quad q \perp E_{ix} . \tag{3.125}$$

From (3.112) one has for $\cos^2 \varphi_1 = 1$

$$\gamma_q = 2\omega \epsilon'' |E_{ix}|^2 \, y(q)/\pi g . \tag{3.126}$$

The dependence $y(q)$ in (3.126) is shown on Fig. 37: In contradistinction from

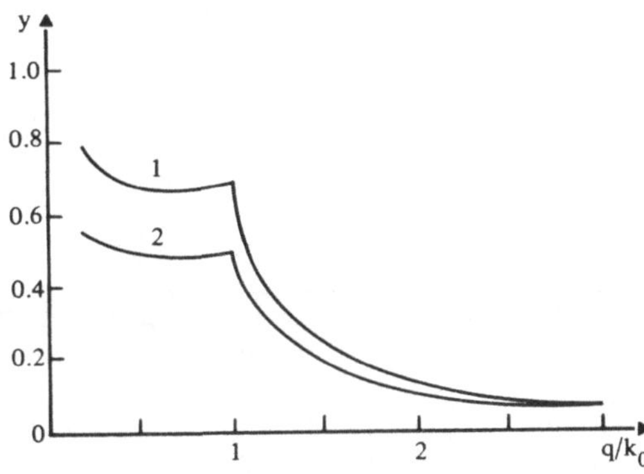

Fig. 37. The characteristic dependency of the growth rate of the transverse gratings on q for $\theta = 0$ in the region $- -1 < \epsilon' < 0$. 1) $\epsilon' = -0.25$; 2) $\epsilon' = -0.81$.

previous case (see Fig. 36) the dependency $y(q)$ does not have the salient maximum.

Thus, when $-1 < \epsilon' < 0$, $|\epsilon'| \gg \epsilon''$ the gratings with orientation (3.125) will grow most rapidly in time, for which the modula of vectors q are in the interval $0 \leqslant q \leqslant k_0$. The generation of these gratings occurs due to the diffracted wave, which decays in the medium ($\gamma_1 = \gamma_{-1} > 0$) and is a bulk wave in vacuum ($\Gamma_1 = \Gamma_{-1}$ is imaginary).

d) $\epsilon' > 1$, $\epsilon' \gg \epsilon''$. One can show in this case, that for any fixed q in (3.112) the inequality $b(q, \epsilon) > a(q, \epsilon)$ takes place. Thus, the maximum of growth rate γ_q is reached for longitudinal gratings for which

$$\sin^2 \varphi_1 = 1, \quad \text{i.e. } q \parallel E_{ix} . \tag{3.127}$$

Inserting $\sin^2 \varphi_1 = 1$ into (3.112), we obtain

$$\gamma_q = \frac{2\omega \epsilon''}{\pi g} \; \frac{\sqrt{\epsilon'} - 1}{\sqrt{\epsilon'} + 1} \; |E_{ix}|^2 \, y(q) \tag{3.128}$$

$$y(q) = \begin{cases} \gamma_1 \Gamma_1 / q(\epsilon' \Gamma_1 + \gamma_1), & \text{if } q \geqslant k_0 \sqrt{\epsilon'}, \\[2em] \dfrac{\gamma_{10} \Gamma_1 \, [\gamma_{10} q - \epsilon' \Gamma_1 (k_0 \sqrt{\epsilon'} - \gamma_{10})}{k_0 \sqrt{\epsilon'} \, (\epsilon' - 1) \, (k_0 \sqrt{\epsilon'} - \gamma_{10}) \, (\epsilon'^2 \Gamma_1^2 + q^2)} & \text{if } k_0 \leqslant q_0 \leqslant k_0 \sqrt{\epsilon'}, \\[2em] \gamma_{10} \Gamma_{10} / k_0 \sqrt{\epsilon'} \, (\epsilon' \Gamma_{10} + \gamma_{10}), & \text{if } 0 \leqslant q \leqslant k_0, \end{cases} \tag{3.129}$$

where γ_{10}, Γ_{10}, γ_1 and Γ_1 are defined as in (3.113). The dependence (3.128) is shown on Fig. 38 for $\epsilon' = 1.5$ and $\epsilon' = 5$.

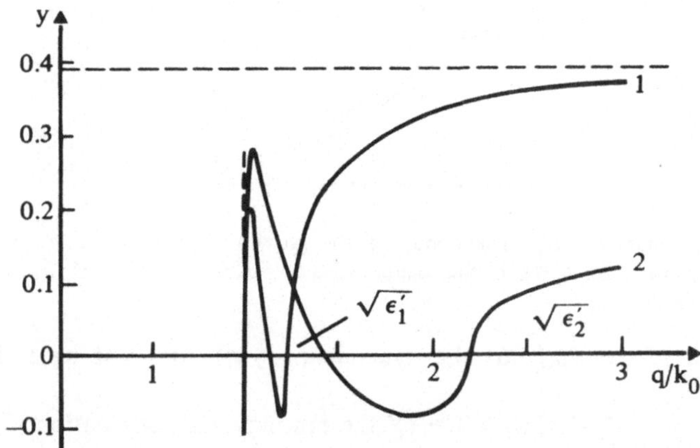

Fig. 38. The characteristic dependence of the growth rate of the longitudinal gratings on q for $\theta = 0$ in the region $\epsilon' > 1$ (according to formula (3.129)). 1) $\epsilon' = 1.5$; 2) $\epsilon' = 5$.

From these figures one can see, that there are regions of the values of q, where $\gamma_q > 0$ as well as regions, where $\gamma_q < 0$. This means, that some Fourie-harmonics of surface relief will grow in time and others will decay under the action of laser light. For $q = 1{,}04\, k_0$ the function $y(q)$ has the sharp maximum, which transforms into absolute one with the growth of ϵ' (see Fig. 38). Thus, the appearence of longitudinal gratings must be observed in the case $\epsilon > 1$. $\epsilon' \gg \epsilon''$, with the parameters

$$\sin^2 \varphi_1 = 1, \qquad \text{i.e.} \quad \mathbf{q} \parallel \mathbf{E}_{ix}, \quad \mathbf{q} \perp \mathbf{k}_t \;;$$

$$\Gamma_1 = 0, \qquad \text{i.e.} \quad q \approx k_0. \tag{3.130}$$

The diffracted waves, leading to the generation of dominant gratings (3.130) with $q \approx k_0$ (but $q > k_0$) are characterized by the imaginary constants $\gamma_1 = \gamma_{-1} = -i\gamma_{10}$ and real constants $\Gamma_1 = \Gamma_{-1} > 0$. Thus, these waves are of the bulk character in medium (see (3.5)), but decay with the distance from the surface into vacuum.

For $\theta \neq 0$, as one can see from (3.111), the maximum of growth rate is reached under the conditions, analogous to (3.130): $\Gamma_1 \approx 0$, $\mathbf{q} \parallel \mathbf{E}_{ix}$, whence it follows that one dominant grating must be generated with the parameters $q = k_0 \cos\theta$; $\mathbf{q} \parallel \mathbf{E}_{ix}$.

3.5.2. p-polarization of the pump wave

Using in (3.110) the equations ((3.72), (3.61)) we obtain the following expression for the complex frequency in the case of p-polarized pump

$$i\Omega = \frac{2\omega \epsilon'' k_z^2 |E_i|^2}{\pi g\, k_0^2\, |k_z \epsilon + i\gamma|^2} \cdot \left[(1 - \epsilon) \cdot \right.$$

$$\tag{3.131}$$

$$\cdot \; \frac{k_s^2 |\gamma|^2 \sin^2 \varphi_1 - |\gamma|^2 \Gamma_1 \gamma_1 + k_1^2 k_t^2 + k_1 k_t (\gamma^* \gamma_1 - \Gamma_1 \gamma) \cos \varphi_1}{(\gamma_1 + \gamma^* + q)\,(\epsilon \Gamma_1 + \gamma_1)} + (1 - \epsilon^*) \cdot$$

$$\cdot \; \left. \frac{k_{-1}^2 |\gamma|^2 \sin^2 \varphi_1 - |\gamma|^2 \Gamma_{-1}^* \gamma_{-1}^* + k_{-1}^2 k_t^2 + k_{-1} k_t (\gamma_{-1}^* \gamma - \Gamma_{-1}^* \gamma^*) \cos \varphi_{-1}}{(\gamma_{-1}^* + \gamma + q)\,(\epsilon^* \Gamma_{-1}^* + \gamma_{-1}^*)} \right] \cdot$$

In the particular case $\theta = 0$, this formula has the form ($\gamma_q = \mathrm{Re}\, i\Omega$):

$$\gamma_q = \frac{2\omega\epsilon''}{\pi g}\, \frac{k_0^2\, |E_i|^2}{|k_0+i\gamma|^2}\, 2\,\mathrm{Re}\left[\frac{\epsilon-1}{\gamma_1+\gamma^*+q}\left(\frac{\gamma_1\Gamma_1\cos^2\varphi_1}{\epsilon\Gamma_1+\gamma_1} - \frac{k_0^2\sin^2\varphi_1}{\Gamma_1+\gamma_1}\right)\right], \qquad (3.132)$$

which coincides with (3.112) with the substitution $\sin^2\varphi_1 \leftrightarrows \cos^2\varphi_1$.

Let us again study seperately four above mentioned regions of the values of ϵ'.

a) Surface polaritos region ($\epsilon' < -1$, $|\epsilon'| \gg \epsilon''$). From (3.131) in this case we obtain

$$\gamma_q = \frac{k_0\,\omega\,\epsilon''\,|E_0|^2\cos^2\theta}{\pi g(|\epsilon'|\cos^2\theta+1)\sqrt{|\epsilon'|}}\, \mathrm{Re}\left(\frac{f(\theta,\varphi_1)}{\Delta k_1 - i\Gamma_p} + \frac{f(\theta,\varphi_{-1})}{\Delta k_{-1} + i\Gamma_p}\right), \qquad (3.133)$$

where $f(\theta,\varphi_a) = \cos^2\varphi_a - \sin\theta\,\cos\varphi_a$.

As in the case of s-polarized pump wave, let us consider two-resonance and one resonance cases.

Two-resonance case. It is realized either at small angles θ, or at $\cos\varphi_a \approx \sin\theta$. However at $\cos\varphi_a = \sin\theta$ in (3.133) we have $f(\theta,\varphi_a) = 0$, and so it is sufficiently to consider only the case of small angles θ. In this case from (3.133) one has ($\theta \approx 0$) at resonance ($\Delta k_a \approx \Gamma_p$) for the growth rate

$$\gamma_q = \frac{|E_0|^2\,\omega}{\pi g}\, 2\cos^2\varphi_1 \qquad (3.134)$$

The modulus and orientation of vector $\mathbf{q}$ are determined by equation (3.115).

One-resonance case. At $\theta \gg a\Gamma_p/4k_0$ for any vectors $\mathbf{q}$ the vector $\mathbf{k}_1$ fulfills the resonance condition (3.114), and the vector $\mathbf{k}_{-1}$ does not. So we can neglect in (3.133) the contribution $f(\theta, \varphi_{-1})$. The analysis of (3.133) shows that for the gratings for which $-1 \leqslant \cos\varphi_1 \leqslant 0$ or $\sin\theta \leqslant \cos\varphi_1 \leqslant 1$ (see Fig. 35b) the growth rate is positive at $\Delta k_1 \approx \Gamma_p > 0$. In this case the time evolution of gratings is described by complex frequency

$$i\Omega = \gamma_q + i\Omega_q$$

$$\Omega_q = \gamma_q = \frac{|E_0|^2\omega|\epsilon'|\cos^2\theta}{\pi g(|\epsilon'|\cos^2\theta+1)}\, |f(\theta,\varphi_1)| \equiv a_2|f(\theta,\varphi)|. \qquad (3.135)$$

At the same time for the gratings for which $0 \leqslant \cos\varphi_1 \leqslant \sin\theta$ the growth rate will be positive, if $\Delta k_1 \approx -\Gamma_p < 0$; the time evolution of these gratings is described by the growth rate (3.135) and $\Omega_q = -\gamma_q$; the modulus and direction of the vector $\mathbf{q}$ are given by (3.115). We note that the growth rate (3.135) ($\theta \neq 0$) at small values of q is approximately two times smaller than the growth rate (3.134). To determine the dominant gratings let us compare the growth rates (3.134), (3.135) at various values of $\cos\varphi_1$. The dependence of growth rate (3.135) ($\theta \neq 0$) on $\cos\varphi_1$ is shown on Fig. 35b. Analysing the results obtained above, we come to the following conclusions:

1) At small θ two almost equally intensive gratings for which $\cos\varphi_1 = \pm 1$ will dominate. The périods of such gratings are determined from (3.115)

$$d_1 = \lambda(1 - \sin\theta), \qquad \mathbf{q} \uparrow\uparrow \mathbf{k}_t$$

$$d_2 = \lambda(1 + \sin\theta), \qquad \mathbf{q} \uparrow\downarrow \mathbf{k}_t \qquad\qquad (3.136)$$

Due to the rather smooth dependence of the growth rate (3.134), (3.135) on $\cos\varphi_1$, the continuum of gratings with parameters near to the parameters, defined by (3.136), must be generated;

2) With the growth of angle θ, the growth rate of grating with period d_1 decreases and that of grating with d_2 increases, so that $\gamma_{q2} > \gamma_{q1}$.

3) When $1 - \sin\theta < \sin^2\theta/4$ (see Fig. 35b), i.e. when $\theta \gtrsim 56^\circ$, besides the grating d_2 with the largest growth rate, another grating d_3 will dominate with

$$\cos\varphi_1 = \sin\theta/2, \qquad \sin\delta_3 = \sqrt{1-\sin^2\theta/4}, \qquad d_3 = \lambda \qquad (3.137)$$

where δ_3 – the angle between $\mathbf{q}$ and $\mathbf{k}_t$ and the growth rate γ_{q3} is determined from (3.135).

4) For $\theta \approx 56^\circ$ two dimensional structures are possible, which are the superposition of the gratings d_2 and d_3.

Let us now go over to the consideration of other regions of values of ϵ. And again as in the case of s-polarization we will consider for the sake of simplicity the normal incidence, using (3.132), and then generalize the results to the case $\theta \neq 0$. Carrying out the analysis analogous to that of previous Sec. 3.5.1. and using the results, represented on Fig. 36–38, one can draw the following conclusions:

b) If $0 < \epsilon' < 1$, $\epsilon' \neq \epsilon''$, the maximum of γ_q for $\theta = 0$ is reached for gratings for which $\sin^2\varphi_1 = 1$ (i.e. $\mathbf{q} \perp \mathbf{E}_{iy}$), $\gamma_1 = 0$ (i.e. $q \approx k_0\sqrt{\epsilon'}$). More general analysis, based on (3.131), shows that for $\theta \neq 0$, one dominant transverse grating is generated with the parameters

c) If $\epsilon' > 1$, $\epsilon' \gg \epsilon''$, then the maximum of γ_q for $\theta = 0$ is reached for the gratings for which $\cos^2 \varphi_1 = 1$ (i.e. $\mathbf{q} \parallel \mathbf{E}_{iy}$), $\Gamma_1 \approx 0$ (i.e. $q \approx k_0$) and for $\theta \neq 0$, according to (3.131), two dominant longitudinal gratings are generated with the parameters $\mathbf{q} \parallel \mathbf{E}_{iy}$, $q = k_0 (1 \pm \sin\theta)$.

The Table VI shows the unified list of above obtained results concerning the periods and orientations of dominant gratings. One can see that at normal incidence the period of longitudinal gratings $d \leqslant \lambda$, and the period of transverse ones $d \geqslant \lambda$.

Table VI. The dependence of the periods and orientations of gratings on $\epsilon = \epsilon' + i\epsilon'' = (n + im)^2$ $(|\epsilon'| \gg \epsilon'')$ [172].

ϵ	$\epsilon' < -1$	$-1 < \epsilon' < 0$	$0 < \epsilon' < 1$	$\epsilon' > 1$
Type of lattice	Normal	Anomalous	Anomalous	Normal
Orientation and period with s polarization	$\mathbf{q} \parallel \mathbf{E}_{ix}$, $\dfrac{\lambda}{(n^{*2} - \sin^2\theta)^{1/2}}$	$\mathbf{q} \perp \mathbf{E}_{ix}$, $0 < q < k_0$	$\mathbf{q} \perp \mathbf{E}_{ix}$, $\dfrac{\lambda}{|(\epsilon')^{1/2} \pm \sin\theta|}$	$\mathbf{q} \parallel \mathbf{E}_{ix}$, $\dfrac{\lambda}{(n^{*2} - \sin^2\theta)^{1/2}}$
Orientation and period with p polarization	$\mathbf{q} \parallel \mathbf{E}_{iy}$, $\dfrac{\lambda}{n^* \pm \sin\theta}$	$\mathbf{q} \perp \mathbf{E}_{iy}$, $0 < q < k_0$	$\mathbf{q} \perp \mathbf{E}_{iy}$, $\dfrac{\lambda}{(\epsilon' - \sin^2\theta)^{1/2}}$	$\mathbf{q} \parallel \mathbf{E}_{iy}$, $\dfrac{\lambda}{n^* \pm \sin\theta}$

$n^{*2} = 1 + (n+m)^2/(m^2+n^2)^2$ for $|\epsilon| \gg 1$. $n^{*2} = |\epsilon'|/(|\epsilon'| - 1) \cong 1$ for $\epsilon' \ll -1$ (sew).

As was mentioned above, all the results for the dependence of the gratings orientations and periods on the value of ϵ, obtained in the case of interferencial evaporation instability, are valid in the case of CW and SAW generation. The detailed theoretical and experimental studies of generation of the longitudinal and transverse gratings in dependence of ϵ, confirming this theory were performed for the case of CW and SPS generation on fused silica in the Ref [267].

3.6. The nonlinear regime of the laser generation of capillar waves and the formation of surface ordered structures

The theory of SPS formation expounded in the previous sections was developed for the nonstationary, linear in the amplitude of the modulation of the relief regime. In the frame of the linear theory (i.e. the calculations of the growth

126

rates) the question about the magnitudes of the stationary amplitudes of SPS remains open. The answer on this question is important not only for the obtaining the resulting picture of the surface relief, but also for the adequate description of the radiation-matter interaction itself. We shall show in Sec. 3.7, that under the growth of the SPS-amplitudes the strong enhancement of the surface absorptivity can take place, which leads to the strong enhacement of the rate of the laser heating of the surface.

As it is follows from the previous considerations, the main mechanism of SPS-formation at relatively high pump intensity is the generation of the capillar waves. In the present section the conlinear regime of the CW generation is considered and the stationary SPS amplitudes are calculated in dependence on q (for simplicity we consider only the thermocapillar mechanism of CW exitation, compare Sec. 3.4). The principal cause of the nonlinearity in this problem is the nonlinearity of diffraction of the light wave on the surface modulation with large amplitudes. As the arbitrary surface relief can be represented by its spatial Fourie-spectrum, the problem of the difraction on the surface roughness can be reduced to the problem of diffraction on the sinusoidal modulation with the amplitudes $\xi_q \equiv 4a_q$. We shall consider here only the intramode nonlinearity, which arises as a result of selfinteraction of the grating q owing to the rescattering on it of the diffracted waves (the electrodynamical intramode nonlinearity).

The problem of the nonlinear diffraction on the sinusoidal modulation of the relief with the large amplitude was analyticaly solved in the work [265] under the conditions $|\epsilon| \gg 1$, $(k_0 \xi_q)^2 \ll n/(m^2 + n^2)$. We shall use the general nonlinear in ξ_q expessions for the light field amplitudes obtained in [265] for the study of the nonlinear regimes of SPS-generation.

The experimental studies of SPS formation [141, 268] shows, that the resulting picture of the surface relief is changed with the increase of the laser intensity and (or) with the increase of the number of the radiation pulses. The one-dimensional gratings are eventually replaced by the hexagonal ones [141, 268]. For the full description of this phenomenon the multimode diffraction theory is needed, which so far has not been developed. However, we shall show, that the ideas, developed in the present section, enable to give the physical mechanism of the hexagonal SPS formation [266].

3.6.1. The diffraction of the light on the periodically modulated surface relief with large amplitude of modulation

The relief of the surface is given as before by the equation (3.2) and the incident wave is given by the formula (3.3). In the case of a large amplitude ξ_q in general case one has for a field outside the medium instead of (3.4)

$$E(x,y,z,t) = E_i \, e^{ik_t y + ik_z z - i\omega t} + \sum_{p=-\infty}^{\infty} E_p' \, e^{ik_p r + \Gamma_p z - i\omega t} + c.c. \qquad (3.138)$$

and for the field inside the medium

$$E(x,y,z,t) = \sum_{p=-\infty}^{\infty} E_p \, e^{ik_p r - \gamma_p z - i\omega t} + c.c. \,, \quad k_p = k_t - pq, \ p=0,\pm 1 \,, \qquad (3.139)$$

where the index p defines the order of the diffraction. Using the boundary conditions (the continuity of the tangential components of the field at the interface $z = \xi(r, t)$ one obtains the infinite, nonhomogeneous system of coupled equation for the amplitudes $E_{px}, E_{pz}, E_{px}', E_{pz}'$ [265]. The analytical solution of this system is, of course, impossible. The situation of interest here, however, is when one or both diffracted waves with $p = \pm 1$ are close to the resonance, (see Sec. 3.4.3), i.e. $k_p \approx k_0$. Confining ourselves by consideration of this case, we retain in the equations (3.139) only the terms with $p = 0, \pm 1$. The effects of the second order diffraction ($p = \pm 2$) are taken into account in the work [300]. Using the results obtained in this resonance approximation in [265], under the conditions

$$|\epsilon| \gg 1, \quad \cos\theta \gg 1/\sqrt{m^2 + n^2} \,,$$

$$k_0^2 \, |\xi_q|^2 \ll n/\sqrt{m^2 + n^2} \,, \qquad (3.140)$$

we have for the field inside the medium, using the notations

$$E_{i0} = \frac{k_0}{k_t} E_{iz}, \quad r_{ax} = \frac{k_{ax}k_z}{k_0^2}, \quad r_{ay} = \frac{k_t - k_{ay}}{k_0}, \quad r_{aa} = \frac{k_a^2 k_t - k_0^2 k_{ay}}{k_0^3},$$

$$T_a \cong k_0^2 \, |\xi_q|^2 \, \frac{\epsilon k_0}{\epsilon \Gamma_a + \gamma} \, \frac{k_0}{k_z}, \quad \Gamma_a^2 = k_a^2 - k_0^2, \quad \gamma = (m - in)k_0$$

the following expression:

$$\vec{E}(\vec{r}, z,t) = e^{-\gamma(z - \xi(\vec{r}, t))} \sum_{a=0,\pm 1} E_a \, e^{i\vec{k}_a \vec{r} - i\omega_a t} + k.c. \,, \qquad (3.141)$$

128

where the field amplitudes are determined by the equations

$$E_{0x} \cong \frac{2k_z}{k_z+i\gamma}\, E_{ix}, \qquad E_{0y} \cong -\frac{k_z}{k_0}\,\frac{2i\gamma}{\epsilon k_z+i\gamma}\, E_{i0}\,,$$

$$E_{ax} = \frac{2i\gamma k_{ax}}{\epsilon\Gamma_a+\gamma}\,\frac{r_{ax}E_{ix}+r_{aa}E_{i0}}{1-i\sum_{p=\pm1}T_p(r_{px}^2+r_{py}r_{pp})}\,\xi_a\,,$$

$$E_{ay} = \frac{2i\gamma k_{ay}}{\epsilon\Gamma_a+\gamma}\,\frac{r_{ax}E_{ix}+r_{aa}E_{i0}}{1-i\sum_{p=\pm1}T_p(r_{px}^2+r_{py}r_{pp})}\,\xi_a\,, \qquad (3.142)$$

Here $\xi_a = \xi_q$ for $a=1$ and $\xi_a = \xi_q^*$ for $a=-1$. The component E_z is small relative
to E_x and E_y and may be neglected. From the expession (3.141) one can see,
that the amplitude of the diffracted fields inside the medium are proportional
to the electromagnetic resonance factor l_a — (3.12). Thus for $|\epsilon| \gg 1$ the ampli-
tuds of the diffracted fields are sharply increased for the gratings with $k_a \approx k_0$.
From (3.142) it is seen also that amplitudes depend on ξ_q in nonlinear manner
(more extensive nonlinear expessions for the fields inside as well as outside of
the medium is given in Ref. [265]). In linear by ξ_q approximations the expessions (3.141) are reduced to the formulae (3.8–3.10). The case of arbitrary
form of the relief is treated in Ref. [214].

3.6.2. The equations for the nonlinear regime of capillar waves generation

As in Sec. 3.4.1 we consider the melted surface of the metal or semiconductor,
supposing that $qh > 1$, where h is the depth of the melt. We shall consider
the surface relief in the single mode approximation, supposing that only one
resonance grating (3.2) is generated (or the continuum of the independent
gratings). The validity of this approximation we shall consider below. Let us
introduce the new reference system x', y', z in which the wave vector q is
directed along the x'-axis. The surface relief in this frame has the form:

$$z = \xi(x', t) = \xi_q \exp(i\Omega_q t - iqx') + c.c.. \qquad (3.143)$$

In the frame x', y', z the temperature distribution in the region $z \geqslant \xi(x', t)$,
arising owing to the action of the laser radiation is determined by the equations

9 Akhmanov, Laser

$$\frac{\partial T}{\partial t} + (\mathbf{v}\nabla)T = \chi \Delta T + \frac{\nu\rho}{c_v}\left(\frac{\partial v_i}{\partial x_k} + \frac{\partial v_k}{\partial x_i}\right)\left(\frac{\partial v_i}{\partial x_k} + \frac{\partial v_k}{\partial x_i}\right) + \frac{1}{4\pi c_v}\mathbf{E}\frac{\partial \mathbf{D}}{\partial t}$$

$$T(x',z,t=0) = T_{in}, \qquad T(x',z=\infty,t) = T_{in}$$

$$\left(\frac{\partial T}{\partial z} - \frac{\partial T}{\partial x'}\frac{\partial \xi}{\partial x}\right)\Bigg|_{z=\xi(x',t)} = 0 \tag{3.144}$$

For the completeness we write down explicitly the dissipative and hydrodynamical nonlinear terms.

The equations for the motion of the liquid are given by the formulae (3.54). The boundary conditions at the interface $z = \xi(x',t)$ have the form

$$\frac{\partial \varphi}{\partial t} + (\mathbf{v}\nabla)\varphi - g\xi(x',t) + 2\nu\frac{\partial v_z}{\partial z} + \sigma_0\frac{\partial \xi}{\partial x'^2} - \nu\left(\frac{\partial v_z}{\partial x'} + \frac{\partial v_x}{\partial z}\frac{\partial \xi}{\partial x'}\right) +$$

$$+ \sigma_{0T}\left(\frac{\partial T}{\partial x'} + \frac{\partial T}{\partial z}\frac{\partial \xi}{\partial x'}\right)\frac{\partial \xi}{\partial x'}\sqrt{1 + \frac{\partial \xi}{\partial x'}^2} = 0$$

$$\nu\left(\frac{\partial v_x}{\partial z} + \frac{\partial v_z}{\partial x'}\right) + \sigma_{0T}\left(\frac{\partial T}{\partial x'} + \frac{\partial T}{\partial z}\frac{\partial \xi}{\partial x'}\right)\sqrt{1 + \frac{\partial \xi}{\partial x'}^2} -$$

$$- \left(\frac{\partial \varphi}{\partial t} + (\mathbf{v}\nabla)\varphi - g\xi + 2\nu\frac{\partial v_x}{\partial x'} + \sigma_0\frac{\partial^2 \xi}{\partial x'^2}\right)\frac{\partial \xi}{\partial x'} = 0. \tag{3.145}$$

Besides at interface one has the kinematic relation

$$v_z\Bigg|_{z=\xi(x',t)} = \frac{d\xi(x',t)}{dt} = \frac{\partial \xi(x',t)}{\partial t} + v_x\frac{\partial \xi(x',t)}{\partial x'} \tag{3.146}$$

The electrodynamical nonlinearity is taken into account through the quantity, proportional to the power density of the thermal sources

$$\frac{1}{4\pi c_v}\mathbf{E}\frac{\partial \mathbf{D}}{\partial t} = \left\{ f_0 + (f_1\xi_q e^{-iqx'+i\Omega qt} + c.c.) + \right.$$

$$\left. + (f_2\xi_q^2 e^{-2iqx'+2i\Omega qt} + c.c.)\; e^{-\gamma_0(z-\xi)}\right\}, \tag{3.147}$$

where $\gamma_0 = \gamma + \gamma^*$ and the amplitudes f_0 and f_1 near the resonance ($k \approx k_0$) are

$$f_0 = \frac{4\omega m n k_0}{\pi c_v (m^2 + n^2)} \left\{ |E_{ix}|^2 \cos^2\theta + |E_{i0}|^2 + \right.$$

$$\left. + \zeta \frac{|r_{ax} E_{ix} + r_{aa} E_{i0}|^2}{x_a^2 + \left[\beta_n + \zeta k_0^2 |\xi_q|^2 \frac{(1 - \sin\theta \cos\varphi_a)^2}{\cos\theta} \right]^2} k_0^2 |\xi_q|^2 \right\},$$

$$f_1 = \frac{4\omega m n k_0}{\pi c_v (m^2 + n^2)} \left\{ \frac{(r_{1x} E_{ix}^* - k_{1y} E_{i0}^* / k_0)(r_{1x} E_{ix} + r_{11} E_{i0})}{x_1 - i \left[\beta_n + \zeta k_0^2 |\xi_q|^2 \frac{(1 - \sin\theta \cos\varphi_1)^2}{\cos\theta} \right]} + \right.$$

$$+ \frac{(r_{-1x} E_{ix} - k_{-1y} E_{i0} / k_0)(r_{-1x} E_{ix}^* + r_{-1-1} E_{i0}^*)}{x_{-1} + i \left[\beta_n + \zeta k_0^2 |\xi_q|^2 \frac{(1 - \sin\theta \cos\varphi_{-1})^2}{\cos\theta} \right]} \qquad (3.148)$$

Here $a = \pm 1$, $\zeta = 2$ for the double-resonance case, $\zeta = 1$ for single resonance case (see Fig. 27). Further, we shall not be interested in the generation of the relief harmonics, so we neglect the term, proportional to f_2 in (3.146).

If one neglects all hydrodynamical nonlinearities, then the system of the equations in the coordinate system x', y', $z' = z - \xi(x', t)$ is essentially simplified. For $z' \geqslant 0$ we have

$$\Delta'\varphi = 0, \quad \frac{\partial A}{\partial t} = \nu \Delta' A, \quad \text{div}' A = 0, \quad v = \text{grad}'\varphi + \text{rot}' A. \quad (3.149)$$

At the interface $z' = 0$ the following boundary conditions must be satisfied

$$v_z = \frac{\partial \xi(x', t)}{\partial t},$$

$$\frac{\partial\varphi}{\partial t} - g\xi(x', t) + 2\nu \frac{\partial v_z}{\partial z'} + \sigma_0 \frac{\partial^2 \xi(x', t)}{\partial x'^2} = 0$$

$$\nu \left(\frac{\partial v_x}{\partial z} + \frac{\partial v_z}{\partial x} \right) + \sigma_{0T} \frac{\partial T(x', z', t)}{\partial x'} = 0. \qquad (3.150)$$

131

The differentiation in (3.150) is carried out by the coordinates x', y', z'.
The problem of the laser heating of the surface $-$ (3.144) is divided into two stages, according to the representation of the temperature distribution as a sum of the spatially uniform and nonuniform distributions

$$T(x', z', t) = T_0(z', t) + T_1(x', z', t) \qquad (3.151)$$

The spatially uniform distribution is determined by the equations

$$\frac{\partial T_0}{\partial t} = \chi \Delta' T_0 + f_0 e^{-\gamma_0 z'}, \quad z' > 0,$$

$$T_0(z', t = 0) = T_{in}, \quad T_0(z' = \infty, t) = T_{in}, \quad \left.\frac{\partial T_0(z', t)}{\partial z'}\right|_{z'=0} = 0 \qquad (3.152)$$

And the spatially nonuniform distribution is found from the solution of the following equations ($z' \geqslant 0$)

$$\frac{\partial T_1}{\partial t} = \chi \Delta' T_1 + \left(\frac{\partial \xi}{\partial t} - \chi \frac{\partial^2 \xi}{\partial x'^2} - v_z\right) \frac{\partial T_0}{\partial z'} +$$

$$+ (f_1 \xi_q e^{-iqx' + i\Omega_q t} + c.c.) e^{-\gamma_0 z'},$$

$$T_1(x', z', t = 0) = 0, \quad T_1(x', z' = \infty, t' = 0), \quad \left.\frac{\partial T_1(x', z', t)}{\partial z'}\right|_{z'=0} = 0 \qquad (3.153)$$

The system equations (3.143), (3.148)–(3.153) are the starting system for the study of the influence of the electrodynamical (diffraction) nonlinearity on the process of the laser generation of the capillar waves on the surface of the melt.

3.6.3. The dispersion equation for determining the stationary amplitudes of laser induced capillar waves

As it is seen from (3.149), (3.150) for describing the hydrodynamical motions in the melt it is sufficient to determine the spatially nonuniform temperature distribution near surface $z' = 0$.

For the times $t \gg z^2/4\chi$, $t \gg 1/\chi\gamma_0^2$ the solution of (3.152) has the form

$$T_0(z', t) = \frac{2f_0}{\gamma_0}\sqrt{\frac{t}{\pi\chi}} - \frac{f_0}{\chi\gamma_0^2}(e^{-\gamma_0 z'} + \gamma_0 z') \qquad (3.154)$$

The distribution of the velocities in the liquid as a solution of the equation (3.149) can be written in the form

$$v_x = (-iq\varphi_q e^{-qz'} + \delta a_q e^{-\delta z'})e^{-iqx' + i\Omega_q t} + \text{c.c.}$$

$$v_y = 0$$

$$v_z = (-iq\varphi_q e^{-qz'} - iqa_q e^{-\delta z'})e^{-iqx' + i\Omega_q t} + \text{c.c.} \qquad (3.155)$$

where ξ_q, a_q, φ_q are the stationary (for $t \to \infty$) amplitudes of the surface relief, vector and scalar potential, correspondingly.
Let us substitute (3.154), (3.155) into the first equation (3.153). The solution of thus obtained equation we seek in the form

$$T_1(x', z', t) = \left\{ c_1 + c_2 e^{-qz'} + c_3 e^{-\delta z'} + c_4 e^{-(\gamma_0 + q)z'} + c_5 e^{-(\gamma_0 + \delta)z'} + \right.$$

$$\left. + c_6 e^{-\gamma_0 z'} + c_z e^{-\gamma_1 z'} \right\} e^{-iqx' + i\Omega_q t} + \text{c.c.} \qquad (3.156)$$

$$\gamma_T^2 = q^2 + i\Omega_q/\chi, \quad \text{Re } \gamma_T > 0 ,$$

which is valid for the times $t > 1/q^2\chi$, $1/\chi|\delta|^2$, $1/\chi\gamma_0^2$, $1/\chi|\gamma_T|^2$. After the substitution of the formula (3.156) in the system (3.153), (3.154), (3.155) we determine all the coefficients c_m, entering (3.156). As a result we obtain, that the thermocapillar force, proportional $(dT_1/dx')_{z=0}$ is the function of the amplitudes ξ_q, φ_q, a_q

$$\left.\frac{dT_1(x', z', t)}{dx'}\right|_{z'=0} = iq e^{-iqx' + i\Omega_q t}\left\{ \frac{f_0\gamma_T\gamma_0 - f_1\gamma_0}{\chi\gamma_0\gamma_T(\gamma_T + \gamma_0)}\xi_q + \right.$$

$$(3.157)$$

$$\left. + \frac{f_0 q\gamma_0}{\chi^2\gamma_0\gamma_T(\gamma_T + q)(\gamma_T + q + \gamma_0)}\varphi_q + \frac{if_0 q\gamma_0}{\chi^2\gamma_0\gamma_T(\gamma_T + \delta)(\gamma_T + \delta + \gamma_0)}a_q \right\}$$

133

Substituting (3.143), (3.155), (3.157) into the system of the boundary conditions (3.151) and eliminating φ_q and a_q, we obtain the dispersional equation. Separating the imaginary and the real part of this equation, we obtain two equations ($t \to \infty$, $\gamma_q = 0$):

$$\omega_q^2 - \Omega_q^2 + 4\nu^2 q^3 \, \mathrm{Re}\,(q-\delta) = \mathrm{Re}\left\{ \left[1 + 2i\nu q(\delta-q)/\Omega_q\right]A_q \right. +$$

$$\left. + \left[\frac{i\omega_q^2}{\Omega_q} - 2\nu q\delta\right]B_q - \left[\frac{i(\omega_q^2-\Omega_q^2)}{\Omega_q} - 2\nu q^2\right]C_q\right\}, \tag{3.158}$$

$$4\nu q^2 \Omega_q + 4\nu^2 q^3 \, \mathrm{Im}\,(q-\delta) = \mathrm{Im}\ \ \left[1 + 2i\nu q(\delta-q)/\Omega_q\right]A_q +$$

$$+ \left[\frac{i\omega_q^2}{\Omega_q} - 2\nu q\delta\right]B_q - \left[\frac{i(\omega_q^2-\Omega_q^2)}{\Omega_q} - 2\nu q^2\right]C_q \ , \tag{3.159}$$

where

$$A_q = |\sigma_{0T}|\left\{ \frac{f_1 q^2}{\chi\gamma_T(\gamma_T+\gamma_0)} - \frac{f_0 q^2}{\chi(\gamma_T+\gamma_0)} \right\} ,$$

$$B_q = |\sigma_{0T}| \frac{f_0 q^2}{\chi^2\gamma_T(\gamma_T+q)(\gamma_T+q+\gamma_0)} \ ,$$

$$C_q = |\sigma_{0T}| \frac{f_0 q^2}{\chi^2\gamma_T(\gamma_T+\delta)(\gamma_T+\delta+\gamma_0)} \ .$$

Here we have taken into account that $\sigma_{0T} < 0$: f_0 and f_1 as functions of $|\xi_q|^2$ are determined by the expressions (3.148). The equations (3.158) and (3.159) determine the frequencies Ω_q and the stationary values of ξ_q.

3.6.4. The stationary frequencies and amplitudes of the laser induced capillar waves

We now analize the dispersional equations (3.158), (3.159), (3.148). Put for definitness $\Omega_q > 0$, and assume that

$$\nu q^2 \ll \omega_q \ll 2\nu q^2, \quad \gamma_0 \gg q, |\gamma_T|, \quad \nu q^2 \ll \Omega_q \ll \chi q^2 \tag{3.160}$$

Then the equations (3.158, 159) are essencially simplified and acquires the form

134

$$\omega_q^2 - \Omega_q^2 = \frac{|\sigma_{0T}|q^2}{\chi\gamma_0} \, \mathrm{Re}\left\{f_1/q - f_0\right\}, \qquad 4\nu q^2 \Omega_q = \tag{3.161}$$

$$= \frac{|\sigma_{0T}|}{\chi\gamma_0}\left\{ \mathrm{Im}\frac{f_1}{q} + \left(\sqrt{\frac{2\nu q^2}{\Omega_q}} - \frac{\Omega_q}{2\chi q^2}\right)\mathrm{Re}\frac{f_1}{q} - \left(\sqrt{\frac{2\nu q^2}{\Omega_q}} - \frac{\omega_q^2}{2\chi q^2}\right)f_0 \right\}.$$

Let us consider separately single-resonance and double-resonance cases.

1. **The single-resonance case (Fig. 27b).** Let one of the waves is close to the resonance ($k_a \approx k_0$, where either $a = 1$, or $a = -1$, see Fig. 27b). In this case from (3.148) it is easely seen that $\mathrm{Re}(f_1/q) \sim \mathrm{Im}(f_1/q)$. Thus in the second equation (3.161), taking into account of the inequalities (3.160), one can neglect two last terms. Assume further, that

$$\frac{|\sigma_{0T}|q^2}{\chi\gamma_0}\,\mathrm{Re}\left(\frac{f_1}{q} - f_0\right) \ll \omega_q^2. \tag{3.162}$$

Then, from the first equation (3.161) we have $\Omega_q \cong \omega_q$. Substituting this value into the second equation (3.161), using there (3.148), we obtain the following result. If the wave with $a = 1$ ($k_1 = k_t - q$) is in resonance, then

$$4\nu q^2 \omega_q = \frac{2\omega\,|\sigma_{0T}|\,q\,n}{\pi c_v \chi (m^2 + n^2)} \cdot \tag{3.163}$$

$$\frac{(r_{1x}E_{ix}^* - \frac{k_{1y}}{k_0}E_{i0}^*)(r_{1x}E_{ix} + r_{11}E_{i0})\,[\beta_n + k_0^2\,|\xi_q|^2\,(1 - \sin\theta\,\cos\varphi_1)^2\cos^{-1}\theta\,]}{x_1^2 + [\beta_n + k_0^2\,|\xi_q|^2\,(1 - \sin\theta\,\cos\varphi_1)^2\,\cos^{-1}\theta\,]^2}$$

If the wave with $a = -1$ ($k_{-1} = k_t + q$) is in resonance, then

$$4\nu q^2 \omega_q = \frac{-2\omega\,|\sigma_{0T}|\,q\,n}{\pi c_v\,(m^2 + n^2)\chi} \cdot \tag{3.164}$$

$$\frac{(r_{-1x}E_{ix} - \frac{k_{-1y}}{k_0}E_{i0})(r_{-1x}E_{ix}^* + r_{-1-1}E_{i0}^*)\,[\beta_n + k_0^2|\xi_q|^2(1 - \sin\theta\,\cos\varphi_{-1})^2\cos^{-1}\theta\,]}{x_{-1}^2 + [\beta_n + k_0^2\,|\xi_q|^2\,(1 - \sin\theta\,\cos\varphi_{-1})^2\,\cos^{-1}\theta\,]^2}$$

From these expressions it is seen, that the amplitudes $|\xi_q|$ will have the gratest values, when $x_a = 0$, i.e. if

$$k_a^2 = k_0^2 + m^2 k_0^2 / (m^2 + n^2)^2 \approx k_0^2 \,.$$

The wave vectors of the resonance capillar waves are determined then by the relations

$$q = k_0 (1 + \sin^2\theta - 2\sin\theta \, \cos\varphi_a)^{1/2} \,. \qquad (3.165)$$

In the following we put in (3.163) $x_1 = 0$ and in (3.164) $x_{-1} = 0$.
a) **The case of s-polarized incident wave** ($E_{ix} \neq 0$, $E_{i0} = 0$). From (3.163), (3.164) it is seen that the positive values $|\xi_q|^2$ are obtained, if $a = 1$. As a result we have from (3.163)

$$k_0^2 |\xi_q|^2 = \frac{n}{m^2 + n^2} \frac{\cos\theta}{(1 - \sin\theta \, \cos\varphi_1)^2} \; (I_{ix} / I_{th}^{(s)} - 1) \,,$$

$$I_{th}^{(s)} = \frac{\omega_q c_v \chi^\nu q}{|\sigma_{0T}| k_0 \cos^2\theta \, \sin^2\varphi_1} \,, \quad k_1 = k_t - q \,, \quad |k_1| \cong k_0 \,. \qquad (3.166)$$

where $I = cE^2/2\pi$. The direction of the propagation of the stationary capillar waves coincides with the direction of q.
Taking into account the conditions (3.140) (the region of validity of the non-linear diffraction theory of Sec. 3.6.1), we can make a conclusion that the formula (3.166) is valid for

$$1 < I_{ix} / I_{th}^{(s)} < \sqrt{m^2 + n^2} \; (1 - \sin\theta \, \cos\varphi_1) \cos\theta \,. \qquad (3.167)$$

It is easy to verify, that the condition (3.162), used for the derivation of (3.166), is fulfilled in the region defined by the inequality (3.167).
From (3.167) it is seen, that for the stationary regime of CW-generation to be reached, it is necessary for intensity of the incident radiation to exceed

specific threshold value. For example, in the case of the melt of germanium $\rho = 5{,}51$ g/cm^3, $\nu = 1{,}55 \cdot 10^{-3}$ cm^2/c, $\chi = 0{,}1$ cm^2/s, $C_v = 2{,}45 \cdot 10^7$ erg/cm^3, $\sigma = 600$ din/cm, $|\sigma_{0T}| = 0.07$ cm^3/s^2, $m = 7{,}44$, $n = 4{,}84$, for $\theta = 0^0$, $\lambda = 2\pi/k_0 = 1{,}06$ μm, according to (3.166), we obtain $I_{th} = 7{,}8 \cdot 10^5$ W/cm^2 ($q \sim k_0$, $\sin^2\varphi_1 \cong 1$).

b) **The case of p-polarized incident wave ($E_{ix} = 0$, $E_{i0} \neq 0$).** In this case from the expressions (3.163), (3.164) for the resonance amplitudes we obtain

$$k_0^2 \, |\xi_q|^2 = \frac{n}{m^2 + n^2} \, \frac{\cos\theta}{(1 - \sin\theta \cos\varphi_a)^2} \, (I_{i0}/I_{th}^{(p)} - 1) \tag{3.168}$$

$$I_{th}^{(p)} = \frac{\omega_q \, c_v \, \chi \nu \, q}{|\sigma_{0T}| \, k_0 \, |\cos^2\varphi_a - \sin\theta \cos\varphi_a|} \, ,$$

$$I_{i0} = \frac{c \, |E_{i0}|^2}{2\pi} \, ,$$

where

$$a = 1, \quad k_1 = k_t - q, \quad \text{if} \;\; \cos\varphi_1 < 0, \;\; \text{or} \;\; \cos\varphi_1 > \sin\theta$$

$$a = -1, \quad k_{-1} = k_t + q, \quad \text{if} \;\; 0 \leqslant \cos\varphi_{-1} \leqslant \sin\theta \, . \tag{3.169}$$

The modulus of vector **q** is determined by the equation (3.165).

2. **The double-resonance case ($k_a \cong k_0$, $a = \pm 1$) (see Fig. 27a).** Then in the expression (3.148) one must put $x_a = x_1 = x_{-1}$, $\cos^2\varphi_a = \cos^2\varphi_1 = \cos^2\varphi_{-1} \cong \sin^2\theta$. Then $\text{Im} \, f_1 = 0$. Assuming that the condition (3.162) is fulfilled, we obtain from (3.161)

$$\Omega_q \cong \omega_q, \quad a_q \equiv \sqrt{2\nu q^2/\omega_q} \, - \omega_q/2\chi q^2 \, ,$$

$$4\nu q^2 \, \omega_q = \frac{|\sigma_{0T}| q^2}{\chi\gamma_0} \, a_q \;\; \text{Re} \left\{ f_1/q - f_0 \right\} . \tag{3.170}$$

As it is seen from (3.170), (3.148) in the double-resonance case for p-polarized incident radiation the quantity $r_{11} \cong k_0(\sin\theta - \cos\varphi_a) \approx 0$.

This leads to the very great value of I_{th} (this also follows from (3.168)). Thus in the double-resonance case it is sufficient to consider only the case of spolarized pumping wave. Then from the expressions (3.170), (3.148) we obtain

$$4 \nu \omega q = \tag{3.171}$$

$$= \frac{2 a_q \omega |\sigma_{0T}| \, n \, |E_{ix}|^2}{\pi c_v \chi k_0 (m^2 + n^2)} \left\{ \frac{2(\chi_a - k_0^2 |\xi_q|^2 \cos\theta) \cos^3\theta}{x_a^2 + [\beta_n + 2k_0^2 |\xi_q|^2 \cos^3\theta]^2} - \cos^2\theta \right\}.$$

From the equation (3.171) one can find, that the maximum values of $|\xi_q|^2$ are reached for

$$x_a = k_0^2 |\xi_q|^2 \cos\theta \pm ((k_0^2 |\xi_q|^2 \cos\theta)^2 +$$

$$+ \, [\beta_n + 2k_0^2 |\xi_q|^2 \cos^3\theta]^2)^{1/2}, \tag{3.172}$$

where the sign $(+)$ corresponds to the case $a_q > 0$ and the sign $(-)$ for $a_q < 0$. Substituting (3.172) into (3.171), we obtain the equation for the stationary amplitude, from which we have

$$k_0 |\xi_q|^2 =$$

$$= \frac{n \, [((\eta \pm 2\cos^2\theta)^2 + 4(\eta^2 - 1) \cos^4\theta)^{1/2} \mp (\eta \pm 2\cos^2\theta)]}{4(m^2 + n^2) \cos^5\theta},$$

$$\eta = I_{ix}/I_{th}^{(s)}, \quad I_{th}^{(s)} = \frac{\omega_q c_v \chi \nu}{|a_q \sigma_{0T}| \cos^3\theta}, \quad q \cong k_0 \cos\theta \tag{3.173}$$

with the upper sign corresponding to $a_q > 0$, and the lower sign to $a_q < 0$.
Let us compare the expressions for the threshold intensities (3.166), (3.165), (3.173) for the single and double-resonance cases. From (3.166), (3.176) it follows, that $I_{th}^{(s)}(166)/I_{th}(173) \sim |a_q| \ll 1$ (See (3.160)). For example, in the germanium melt for $\lambda = 1 \ \mu$m and $\theta = 0$, we have $a_q = 0{,}06$. Thus, for a given value of the pump intensity the stationary amplitudes of the single-resonance capillary waves are greater than the stationary amlitudes of double-resonance capillary waves.
In Fig. 39 the dependence of the stationary amplitudes $b_q = k_0 |\xi_q| \sqrt{(m^2+n^2)/n}$ of the resonance capillary waves for the given intensity of s-polarized incident wave on the modulus and orientation of the vector q (see Fig. 27 and (3.175)) is given. The curves are calculated with the help of the formula (3.166), (3.173).

138

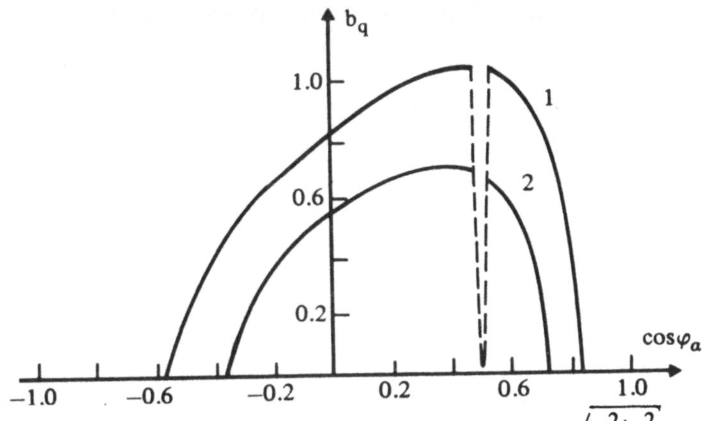

Fig. 39. The dependence of the stationary amplitudes $b_q = k_0 |\xi_q| \sqrt{\dfrac{m^2+n^2}{n}}$ of the CW in germanium on φ_a for s-polarization of the pump wave (according to the formula (3.166), (3.173)). 1) $\theta = 30^0$, $N\cos^2\theta = 2$; 2) $\theta = 30^0$, $N = 2$. $N \equiv I_{ix} |\sigma_0 T|/\omega_q c_v \chi \nu$..

The analogous dependency for p-polarized incident wave (according to the formula (3.178) for $N = I_{i0} |\sigma_0 T|/\omega_q c_v \chi \nu = 2$, $\theta = 30^0$) is given in Fig. 40. As it is seen from Fig. 39, 40 for the intensities, exceeding I_{th}, the continuum of the periodical structures is generated, the width of this continuum being increased with the increase of the intensity of the incident radiation.

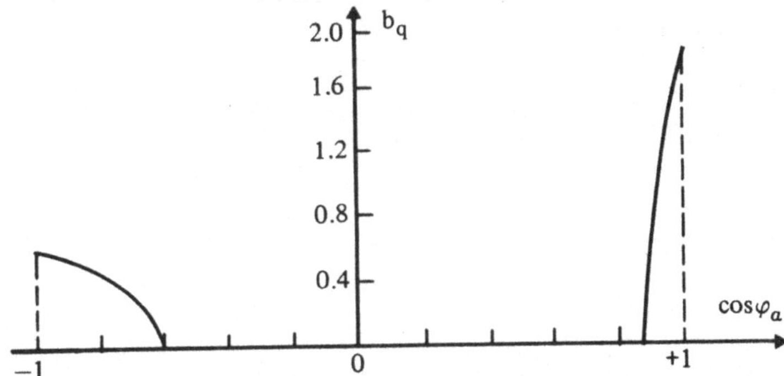

Fig. 40. The dependence of the stationary amplitude b_q of the CW in germanium on φ_a for p-polarization of the pump wave (according to the formula (3.168). $N \equiv I_{i0} |\sigma_0 T| /\omega_q c_v \chi \nu = 2$, $\theta = 30^0$.

In the case of p-polarized radiation (Fig. 40) the capillar waves, corresponding to $\cos \varphi_a = \pm 1$ have the biggest amplitudes. The parameters of these waves are the same as for the waves arisen in the initial nonstationary generation regime, However, in contradistinction to the nonstationary regime, in the stationary regime the amplitude, corresponding to $\cos \varphi_a = 1$ is bigger than the stationary amplitude, corresponding $\cos \varphi_a = -1$. In some experiments the domination of the structures with $\cos \varphi_a = +1$ was observed, while in another experiments-domination of structures with $\cos \varphi_a = -1$ [141, 269] was recorded. From the view-point of the present theory this can be explained by the realization of either stationary, or nonstationary regimes.

For s-polarized incident radiation, as it is seen from Fig. 39 the broad Fourie-spectrum of the wave vectors with the width Δk_s is generated with the narrow and deep minimum in the vicitity of $\cos \varphi_a = \sin \theta$. In the absence of this minimum the resulting picture of the surface relief, corresponding to such a spectrum, would be the localized pit of the width $\Delta \chi \sim (\Delta k_s)^{-1}$. The minimum can be obtained by the compensation of on Fourie-harmonic with $\cos \varphi_a = \sin \theta$ in this spectrum. Thus the resulting picture of the surface relief, corresponding to the Fourie-spectrum of Fig. 39, is the superposition of the localized pit and the grating with $\cos \varphi_a = \sin \theta$, i.e. the grating with the parameters (3.94).

Thus, in stationary regime with the single mode approximation, when the intensity slightly exceeds the threshold value, the same structure dominate, which prevails in the nonstationary regime. The amplitudes of the stationary structures depend on the intensity of the pump (see (3.173), (3.168)).

3.6.5. The multimode regime of capillar wave generation and the formation of the hexagonal surface structures

Let us discuss the region of validity of the obtained results. Firstly, owing to the limitations (3.140) of the nonlinear theory of diffraction, the expressions for the stationary amplitudes (3.173), (3.168, 166) are valid only for small exceeding of the intensity of the incident radiation above the threshold value (see (3.167)). Besides, one must keep in mind that the nonlinear theory of diffraction (Sec. 3.6.1) was developed, assuming that there is only one capillar wave or grating (the single-mode approximation). The nonlinear dependence of the amplitudes of the diffracted waves on ξ_q arises owing to the multiple scattering of the waves, diffracted on the given grating, on the same grating (intramode rescattering). But, as was pointed out earlier, the continuum of the resonance gratings is generated simultaneously under the action of the laser radiation (for which the ends of vectors k_a lie on the circle with the radius $\cong k_0$, see Fig. 27). This means, that, along with taking into account the intramode rescattering, one must, in general, take into account the intermode rescattering, in which the electromagnetic waves, diffracted on some grating are

140

rescattering on another resonance grating. At present there is no multimode nonlinear theory of difraction. Nevertheless we shall discuss below some ideas, which enables determining the condition of validity of single-mode consideration and establishing what periodic relief structures can be formed in multimode regimes.

Let us consider, firstly, the case, when there is only one surface grating. The incident radiation field $\mathbf{E}_i$ with the projection of the wave vector $\mathbf{k}_t$, is scattered on the grating with the wave vector $\mathbf{q}_1$, with the creation of the diffracted wave $\mathbf{E}_1$ with the wave vector $\mathbf{k}_1$ ($\mathbf{k}_t - \mathbf{q}_1 = \mathbf{k}_1$). If $|\mathbf{k}_1| \cong k_0$, the scattering proceeds in the resonant manner, if on the contrary $|\mathbf{k}_1| \neq k_0$, the scattering is nonresonant. For $|\epsilon| \gg 1$ the effectiveness of the resonant scattering $E_1/E_i = = (m^2+n^2)\, k_0\, \xi_{q1}/n$ exceeds that of the nonresonant one $E_1/E_i \sim k_0 \xi_{q1}$ (see Sec. 3.6.1). The diffracted field $\mathbf{k}_1$ can be rescattered on the grating $\mathbf{q}_1$ into zero mode ($\mathbf{k}_1 + \mathbf{q}_1 = \mathbf{k}_t$, $\mathbf{k}_t \neq k_0$), which is the nonresonant process with the effectiveness $\sim k_0 \xi_{q1}^{*}$. This rescattered zero mode is scattered on the grating $\mathbf{q}_1$ into the wave $\mathbf{k}_1 (\mathbf{k}_t - \mathbf{q}_1 = \mathbf{k}_1)$ with the effectiveness $\sim (m^2 + n^2)k_0\, \xi_{q1}/n$, if the wave $\mathbf{k}_1$ is in resonance, or $\sim k_0 \xi_{q1}$, if it is nonresonant. As a result, if the grating $\mathbf{q}_1$ is in resonance ($|\mathbf{k}_1| \sim k_0$), then we get

$$
E_1 / E_i \sim A\frac{m^2+n^2}{n}(k_0 \xi_{q1}) + B\left(\frac{m^2+n^2}{n}\right)^2 k_0^3 |\xi_{q1}|^2\, \xi_{q1}\,, \qquad (3.174)
$$

where A and B are constants of the same order of magnitude. It is this intramode rescatterings are taken into account in the nonlinear theory of the diffraction, developed in Sec. 3.6.1.

Let now the surface relief be formed by the set of the resonant gratings and the grating $\mathbf{q}_1$ is the resonant one. This means, that two arbitrary gratings from this set, for example, the grating with the vectors $\mathbf{q}_1$ and $\mathbf{q}_2$ satisfy the relations

$$
\mathbf{k}_t - \mathbf{q}_1 = \mathbf{k}_1\,, \qquad |\mathbf{k}_1| \cong k_0
$$

$$
\mathbf{k}_t - \mathbf{q}_2 = \mathbf{k}_2\,, \qquad |\mathbf{k}_2| \cong k_0\,. \qquad (3.175)
$$

Then, besides the intermode channel of rescattering the complementary intermode channel appears. Namely, the diffracted wave $\mathbf{k}_1$ (see Fig. 41) arisen on the grating $\mathbf{q}_1$, can be rescattered on

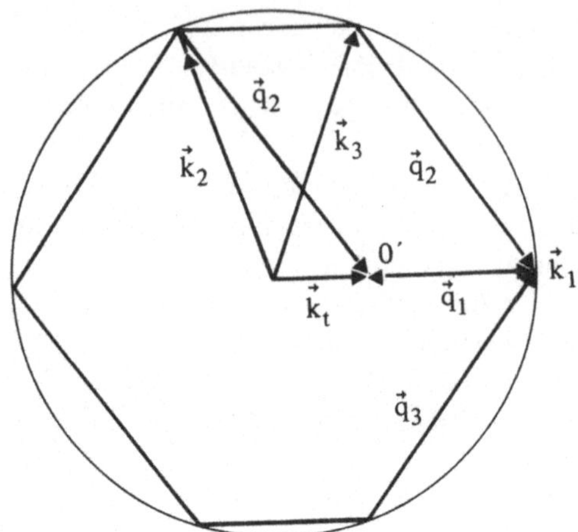

Fig. 41. The kinematics of three gratings q_1, q_2, q_3 interaction through the diffracted field, leading to the hexagonal SPS formation.

the grating q_2 into the wave with the wave vector $k_3 (k - q_2 = k_3)$, which, rescattering on the same grating q_2 $(k_3 + q_2 = k_1)$, gives rise to the additive contribution to the field E_1. This channel is a resonant one, if the following condition is hold

$$k_1 - q_2 = k_3 , \qquad |k_3| \cong k_0 \tag{3.176}$$

The analysis of the relations (3.175), (3.176) shows, that for the dominant resonant grating q_1 from the set of the resonant gratings there are only two gratings q_2, q_3, which yields the resonant intermode scattering channel. The orientation and the magnitudes of the vectors q_2, q_3 are shown in the Fig. 41.

Thus, when the relief of the surface is formed by the resonant gratings only, the amplitude of the field with the wave vector k_1 with taking into account of intramode and intermode resonant channels, can be written in the form

$$E_1/E_i = A\frac{(m^2+n^2)}{n} k_0 \xi_{q1} + B\left(\frac{m^2+n^2}{n}\right)^2 k_0^3 |\xi_{q1}|^2 \xi_{q1} +$$

$$+ C\left(\frac{m^2+n^2}{n}\right)^3 k_0^3 (|\xi_{q2}|^2 + |\xi_{q3}|^2) \xi_{q1} , \tag{3.177}$$

142

where A, B, C are constant of the same order of magnitude. Comparing the expressions (3.177), (3.174) one can draw the following conclusion. If the amplitude of the dominant grating $|\xi_{q1}|$ is such that the inequality

$$|\xi_{q1}|^2 > (|\xi_{q2}|^2 + |\xi_{q3}|^2)(m^2 + n^2)/n \qquad (3.178)$$

holds, then one may neglect the intermode scattering channel and confine oneself, as it was done in this section, with single-mode approximation. On the linear stage of the temporal evolution we have $\xi_{q1} = \xi_0 \exp(\gamma_{q1} t)$, $\xi_{q2} = \xi_0 \exp(\gamma_{q2} t)$, where ξ_0 is the initial amplitude of the relief, $\gamma_{q1} = \gamma_{q1}^0 (I/I_{q1th}^{(s,p)} - 1)$, $\gamma_{q2} = \gamma_{q2}^0 (I/I_{q2th}^{(s,p)} - 1)$. The threshold intensities $I_{qth}^{(s,p)}$ are given by the formula (3.168), (3.166) and the growth rate $\gamma_q^{(0)}$ were calculated in Sec. 3.4.3. The stationary value of the amplitude of the dominant mode ξ_{q1st} (see (3.166), (3.168)) is reached in a time $t_{st} = \gamma_{q1}^{-1} \ln(\frac{\xi_{q1st}}{\xi_0})$. Using these values of the amplitudes in (3.178), and taking into account, that $k_0 \xi_{q1st} = (n/(m^2+n^2))^{1/2} \cdot (I/I_{q1th}^{(s,p)} - 1)^{1/2}$ (see (3.168), (3.166)), one can write down the condition of validity of the single mode approximation in the form

$$(I/I_{q1th}^{(s)} - 1) > (k_0 \xi_0)^2 \left(\frac{m^2 + n^2}{n}\right)^{\frac{2-a}{1-a}} \equiv b,$$

$$a = \frac{\gamma_{q2}}{\gamma_{q1}} = \frac{\gamma_{q2}^{(0)}(I/I_{q2th}^{(s,p)} - 1)}{\gamma_{q1}^{(0)}(I/I_{q1th}^{(s,p)} - 1)} \qquad (3.179)$$

with $\gamma_{q2} = 0$, for $I < I_{q2th}^{(s,p)}$. Because the mode q_1 is assumed to be dominant, the inequalities $\gamma_{q1} > \gamma_{q2}$ and $a < 1$ are always satisfied. If the condition (3.179) is not met, then the one must take into account the intermode wave interactions.

From (3.179) we obtain, that for $I_{q1th}^{(s,p)} < I < I_{q2th}^{(s,p)}$ ($a = 0$) and small initial values of ξ_0 (for germanium we have $I/I_{q1th}^{(s,p)} - 1 > 0$, for $k_0 \xi_0 = 0,01$) the single mode approximation is valid. For sufficiently high intensities ($I > I_{q2th}^{(s,p)}$), $a \to \gamma_{q2}^{(0)}/\gamma_{q1}^{(0)} =$ Const. Thus for the intensities $I > I_{q1th}^{(s,p)}(b + 1) \equiv d_{th}$ the single mode approximation also holds. In the intermediate region of the intensities $I_{q2th}^{(s,p)} < I < d_{th}$. The intermode scattering must be taken into account and the generation of multimode (for example, hexagonal) structures will take place. All the above considerations are valid for the times $t < t_{st}$. For the times $t \gg t_{st}$. when the dominant grating is stabilized, but the other gratings continue to grow $(I > I^{(s,p)}_{q\,2th})$ the generation of multimode structures will occur.

3.7. The self-induced strong enhancement of the material absorptivity under the surface ripples generation.

In studying process of strong laser beam-matter interaction one usually neglects the influence of SPS-generation on the process of the laser energy absorption and the heating of the surface. Then for the temperature of the surface the following expession is used

$$T_0(t) = \frac{2\,I_i}{c_v}\,A\,\cos\theta\,\sqrt{t/\pi\chi}\,,\tag{3.180}$$

where A is the surface absorptivity. For the flat surface $(A \equiv \bar{A})$ of the metals in the optical region $\bar{A} \ll 1$ $(\bar{A} \sim 10^{-1} - 10^{-2})$. We will show in this section that the rough surface with the specific parameters of the roughness (depending on the laser wavelength, polarization and θ) of highly reflecting material $(\bar{A} \ll 1)$ becomes highly absorptive $(A \equiv \tilde{A} \sim 1)$, which leads to the anomalously rapid heating of such surface.

The physical picture of anomalously high rate of heating of rough surface consists in the following. The surface relief can be represented by a set of spatiall Fourie-components. The field inside the medium consists of the sum of the transmitted wave and the waves, diffracted on the modulations of the surface relief. The amplitudes of the resonant diffracted waves already at relatively small amplitudes of the surface roughness $(a_q \sim 0{,}01\,\lambda$, see Sec. 3.2) can be comparable and essencially exceeding the amplitude of the transmitted wave, which determins the absorptivity for the flat surface. This increase of the field inside the medium gives rise to the strong enhancement of the absorptivity.

The resonant Forier-components of the surface relief can be generated on the initially flat surface owing to the SPS-generation. (Sec. 3.4.3). Then the temporal exponential growth of these amplitudes gives rise to the exponential growth of the laser energy absorption and rapid heating of the surface.

144

3.7.1. The dynamics of the laser heating of the surface with the periodically modulated relief.

Let the laser wave (3.3) be incident on the surface with the relief (3.2). The field inside the medium is given then by the expressions (3.141). For the description of the heating of the medium by laser radiation we write down the expression for power density of optical absorption $Q = E\, \partial\, D/4\pi\partial t$, using the formula (3.141), (3.142):

$$Q/c_v = \left\{ f_0 + [f_1 \exp(-iqr) + c.c.] + [f_2 \exp(-2iqr) + c.c.] \right\} e^{-\gamma(z-\xi(x,y))}, \tag{3.181}$$

where $\gamma = \gamma_0 + \gamma_0^* \cong 2mk_0$ (see Appendix, p. 201)

$$f_0 = \frac{\omega\epsilon''}{2\pi c_v} \sum_{a=0,\pm 1} (|\varphi_a{}^*|^2 + |\psi_a|^2), \qquad f_2 = \frac{\omega\epsilon''}{2\pi c_v} (\varphi_1 \varphi_{-1}^* + \psi_1 \psi_{-1}^*) e^{-2i\varphi}$$

$$f_1 = \frac{\omega\epsilon''}{2\pi c_v} (\varphi_0{}^* \varphi_1 + \varphi_0 \varphi_{-1}^* + \psi_0{}^* \psi_1 + \psi_0 \psi_{-1}^*) e^{-i\varphi}.$$

Thus, the problem of the medium heating may be formulated with the help of the following equations:

$$\frac{\partial T}{\partial t} = \chi\Delta T + Q/c_v, \tag{3.182}$$

$$T(x, y, z, t = 0) = T_{in}, \qquad T(x, y, z = \infty, t) = T_{in},$$

$$\left(\frac{\partial T}{\partial z} - \frac{\partial T}{\partial x}\frac{\partial \xi(x,y)}{\partial x} - \frac{\partial T}{\partial y}\frac{\partial \xi(x,y)}{\partial y} \right)_{z=\xi(x,y)} = 0$$

Let us introduce the substitution

$$z' = z - \xi(x, y) \tag{3.183}$$

and use the condition $\partial^2 T/\partial z'^2 > \partial^2 T/\partial x\,\partial z'$, $\dfrac{\partial^2 T}{\partial y\partial x'}$, (the fulfilment of which can be readily checked in the end of the calculations). Then under the condition $k_0 a_q \ll 1$ we have

10 Akhmanov, Laser

$$\Delta T \approx \frac{\partial^2 T}{\partial x^2} + \frac{\partial^2 T}{\partial y^2} + \frac{\partial^2 T}{\partial z'^2} \equiv \Delta' T .$$

Taking into account these remarks, we transform the problem (3.182) (3.181) to the form

$$\frac{\partial T}{\partial t} = \chi \Delta' T + \left\{ f_0 + (f_1 e^{-iqr} + \text{c.c.}) + (f_2 e^{-2iqr} + \text{c.c.}) \right\} e^{-\gamma z'}, \qquad (3.184)$$

$$T(x, y, z', t=0) = T_{in}, \quad T(x, y, z'=\infty, t) = T_H, \quad \left. \frac{\partial T}{\partial z'} \right|_{z'=0} = 0$$

At times $t \gg (z')^2/4\chi$, $t \gg 1/\chi\gamma^2$, $1/\chi q^2$ (3.185)
the solution of (3.184) has the form

$$T(x, y, z', t) - T_{in} = \frac{2f_0}{\gamma} \sqrt{\frac{t}{\pi\chi}} - \frac{f_0}{\chi\gamma^2}(e^{-\gamma z'} + \gamma z') + \qquad (3.186)$$

$$+ \quad (c_1 e^{-\gamma z'} + c_2 e^{-qz'}) e^{-iqr} + (c_3 e^{-\gamma z'} + c_4 e^{-2qz'}) e^{-2iqr} + \text{c.c.} \quad ,$$

where $\quad c_1 = f_1/\chi(q^2 - \gamma^2), \quad c_2 = -\gamma f_1/\chi q(q^2 - \gamma^2),$

$$c_3 = f_2/\chi(4q^2 - \gamma^2), \quad c_4 = -\gamma f_2/2\chi q(4q^2 - \gamma^2) .$$

That part of the solution (3.186), which for z' =const does not depend on x, y, can be called the spatially uniform solution. As can be seen from (3.186) the first term of the spatially uniform solution for the times (3.185) is much bigger then the second one. As we will be interested in the spatially uniform heating only, we write down the expression for the temperature in the form

$$T(x, y, z', t) - T_{in} = \frac{2f_0}{\gamma} \sqrt{\frac{t}{\pi\chi}} , \qquad (3.187)$$

where, with taking into account (3.141), (3.181) we have

$$f_0 \approx \frac{\omega\epsilon''}{2\pi c_v} \left\{ \frac{4k_z^2}{|k_z + i\gamma|^2} |E_{ix}|^2 + \frac{k_z^2}{k_0^2} \frac{4|\gamma_0|^2}{|\epsilon k_2 + i\gamma_0|^2} |E_{i0}|^2 + \right. \qquad (3.188)$$

$$\left. + \frac{4k_0^2}{|\gamma_0|^2} \sum_{a=\pm1} \left| \frac{\epsilon k_0}{\epsilon\Gamma_a + \gamma_a} \right|^2 \frac{|r_{ax} E_{ix} + r_{ay} E_{i0}|^2 (k_0 a_q)^2}{\left| 1 - i(k_0 a_q)^2 \frac{k_0}{k_z} \sum_{p=\pm1} \frac{\epsilon k_0}{\epsilon\Gamma_p + \gamma_0}(r_{px}^2 + r_{py}^2) \right|^2} \right\} .$$

3.7.2. The enhanced absorptivity and rapid heating of the surface for s-polarization of the pumpwave

Let us determine the rate of surface heating in the case $E_{ix} \neq 0$, $E_{i0} = 0$ with taking into account (3.84), (3.85), we obtain from (3.188) for $k_a > k_0$.

$$f_0 = \frac{k_0 I_{ix}}{c_v} \frac{8mn}{m^2 + n^2} \cos^2\theta \cdot$$

$$\cdot \left\{ 1 + \sum_{a=\pm 1} \frac{(k_0 a_q)^2 \sin^2\varphi_a / (x_a^2 + \beta_n^2)}{\left[1 + (k_0 a_q)^2 \sum_{p=\pm 1} \frac{\beta_n}{x_p^2 + \beta_n^2} \frac{(1 - \sin\theta \cos\varphi_p)^2}{\cos\theta} \right]^2 +} \right.$$

$$\left. \frac{}{+ \left[(k_0 a_q)^2 \sum_{p=\pm 1} \frac{x_p}{x_p^2 + \beta_n^2} \cdot \frac{(1 - \sin\theta \cos\varphi_\theta)^2}{\cos\theta} \right]^2} \right\}. \tag{3.189}$$

where $I_{ix} = c E_{ix}^2 / 2\pi$, and the angles φ_a are shown on Fig. 27. The analysis of the expression (3.189) shows that the second term in the square brackets reaches the maximum value in the two-resonance case (see Fig. 27), when $x_1 = x_{-1} = 0$, $\cos\varphi_p = \sin\theta$ for $p = \pm 1$. This corresponds to the grating with the parameters

$$q \parallel E_{ix} \ , \quad q \perp k_t \ , \quad q = k_0 (n^{*2} - \sin^2\theta)^{1/2} , \quad n^{*2} = 1 + m^2 / (m^2 + n^2)^2 . \tag{3.190}$$

Thus, under the action of s-polarized radiation on the surface with the relief, determined by the formula (3.2), (3.190), we obtain for the spatially uniform temperature from (3.187), (3.189) the following expression

$$T(x, y, z, t) - T_{in} = \frac{2 I_{ix}}{c} \tilde{A}_s(a_q) \cos\theta \sqrt{\frac{t}{\pi \chi}}, \tag{3.190}$$

$$\tilde{A}_s(a_q) = \frac{4n}{m^2 + n^2} \cos\theta \left\{ 1 + \frac{2 (k_0 a_q)^2 \cos^2\theta}{[\beta_n + 2(k_0 a_q)^2 \cos^3\theta]^2} \right\},$$

where $\quad \beta_n = n / (m^2 + n^2) .$

The comparison of (3.190) with (3.180) shows, that the quantity $\tilde{A}_s(a_q)$ is the absorptivity of the surface with the relief given by the formula (3.2) (3.190).

147

For $a_q = 0$ the quantity A_s coincides with the absorptivity of the medium with the flat interface. The maximum value of $A_s(a_q)$ is reached for the amplitude of the grating, determined by the relation

$$(k_0 a_{s0})^2 \equiv (k_0 a_q)^2 = n/2(m^2+n^2)\cos^3\theta , \qquad (3.191)$$

i.e. in the conditions, when the full disappearance of the specular reflected from the surface component of the field occurs (see [265]). For $a_q = a_{s0}$ the absorptivity

$$\tilde{A}_s(a_q) \cong 1. \qquad (3.192)$$

The obtained results anable drawing the following conclusion. The highly reflective materials ($|\epsilon| \gg 1$, the flat interface) strongly absorb s-polarized radiation, if in the spatial Fourie-spectrum of the surface relief there are the gratings with the parameters, determined by the formula (3.189), (3.191).

3.7.3. The enhanced absorptivity and rapid heating of the surface by p-polarized pumpwave

Let the incident field be polarized in the plane of incidence ($E_{ix} = 0$, $E_{i0} \neq 0$). We consider the case $\theta \neq 0$, because for $\theta = 0^o$ the results are the same as for s-polarized pump wave incident at $\theta = 0$. Besides, for p-polarization only single resonance case is important (see Fig. 27b), because in the expression (3.188), for the case of geometry shown on Fig. 27a, the quantity $r_{ay} = 0$. From (3.188) and (3.84), (3.85) we obtain ($I_{i0} = \dfrac{cE_{i0}^2}{2\pi}$, $k_a \gg k_0$)

$$f_0 = \frac{k_0 I_{i0}}{c_v}\frac{8mn}{m^2+n^2} \cdot$$

$$\cdot \left\{ 1 + \frac{(k_0 a_q)^2 (\sin\theta - \cos\varphi_a)^2 / (\chi_a^2+\beta_n^2)}{\left[1 + (k_0 a_q)^2 \dfrac{\beta_n}{\chi_a^2+\beta_n^2}\dfrac{(1-\sin\theta\cos\varphi_a)^2}{\cos\theta}\right]^2 + \left[(k_0 a_q)^2 \dfrac{\chi_a}{\chi_a^2+\beta_n^2}\dfrac{(1-\sin\theta\cos\varphi_a)^2}{\cos\theta}\right]^2} \right\}, \qquad (3.193)$$

where $a = 1$, or $a = -1$. The second term of this expression, linked with the surface roughness, reaches the maximum value at $x_a = 0$, $\cos\varphi_a = \pm 1$, i. e. for the gratings with the following parameters

$$\mathbf{q} \parallel \mathbf{E}_{iy}, \quad (\mathbf{q} \parallel \mathbf{k}_t), \quad q = k_0(n^* \pm \sin\theta), \quad n^{*2} = 1 + \frac{m^2}{(m^2+n^2)^2}. \tag{3.194}$$

The spatially uniform heating of the surface with the relief (3.2), (3.190), according to (3.186) is described by the expression

$$T(x, y, z, t) - T_{in} = \frac{2I_{i0}}{c_v} \widetilde{A}_p(a_q)\cos\theta \sqrt{\frac{t}{\pi\chi}}, \tag{3.195}$$

where the absorptivity of the surface with the relief, determined by (3.2), (3.190), in the case of p-polarized radiation has the form

$$\widetilde{A}_q(a_q) = \frac{4n}{(m^2+n^2)\cos\theta}\left\{1 + \frac{(1 \pm \sin\theta)^2\,(k_0a_q)^2}{[\beta_n + (k_0a_q)^2(1 \pm \sin\theta)^2/\cos\theta]^2}\right\} \tag{3.196}$$

The maximum value of $\widetilde{A}(q)$, and consequently, the maximum heating rate are reached, when

$$(k_0a_{p0})^2 \equiv (k_0a_q)^2 = \frac{n}{m^2+n^2}\,\frac{\cos\theta}{(1 \pm \sin\theta)^2}. \tag{3.197}$$

For $a_q = a_{p0}$ we have then

$$\widetilde{A}_p(a_q) \approx 1 \tag{3.198}$$

As it was shown in Ref. [265, 300], the grating with the parameters (3.197), (3.194) have that property, that under the diffraction on it of p-polarized radiation, the full disappearance of the specular reflected vacuum component of the field occurs. From (3.195)–(3.198) it is seen, that for the parameters of the relief, given by the formula (3.2), (3.197), (3.194), the strong enhancement of the heating rate in comparison to the case of the flat surface occurs (the enhancement factor being equal to $(m^2 + n^2)\cos\theta/4n$).

3.7.4. The discussion and the comparison with the experiment

The anomalously high heating rate can be reached under the action of the power-full laser radiation on the initially flat surface. As it was shown in Sec. 3.6 the flat interface becomes unstable, if the intensity of the incident radiation I_i exceeds the critical value I_{th}. For $I > I_{th}$ the amplitudes of the gratings grow exponentially with the growth rate γ_q:

$$\xi_q = \xi_{q0} \exp(\gamma_q t). \tag{3.199}$$

In Sec. 3.4.3 it was shown, that the maximum growth rate has the grating with the parameters, having the values very close to that given by the formula (3.189, 194), namely

$$q \parallel E_{ix}, \quad q = k_0 (n^{*2} - \sin^2\theta)^{1/2},$$

$$n^{*2} = 1 + (m+n)^2 / (m^2 + n^2)^2 \tag{3.200}$$

for s-polarization of the incident wave, and

$$q \parallel E_{iy}, \quad q = k_0 (n^* \pm \sin\theta)^{1/2},$$

$$n^{*2} = 1 + (m+n)^2 / (m^2 + n^2) \tag{3.201}$$

for p-polarization of the pumping wave. This means, that for a sufficiently high pulse energy (such that $\gamma_q \tau_p > 1$) the growth of the amplitudes (3.199) will be accompanied by the sharp increase of the absorptivity and the heating rate (see (3.190), (3.194), (3.195)).
Such picture of the interaction of laser radiation with the surface is in accordance with the experiments [270–273, 188], in which the changes of the absorptivity and reflectivity of the metals during the action of the laser pulse in dependence on the intensity of the pumping wave were studied. In particular in Ref. [188] it was shown, that along with the growth of the resonant amplitudes a_q the absorptivity of the surface of the liquid Hg was sharply increased. For the intensities of the order of $I_i \cong 4,3 \cdot 10^7$ W/cm^2 ($\lambda = 10,6$ μm, the laser pulse duration $\tau_p = 150$ ns, $\theta = 0^0$) the growth of the resonant amplitude of the grating $4a_q$ up to the value of order of 2 μm was accompanied by the complete absorption of the pumping wave and by the rapid reaching of the plasma formation regime. If we put $4\, n/(m^2 + n^2) = 0,15$ [37], then, according to the formula (3.191), the optimum relief depth, for which the 100% absorp-

tion must take place, is equal to $4a_q = 1$ μm. This estimate is fairly well corresponds to the result of the experiment [188].

Another class of the experiments with which the theory can be compared, consists in the investigation of the absorptivity under the action of p-polarized radiation (λ = 10,6 μm) on the Al-grating (with the period d = 1,6 μm and the depth of the modulation $4a_q$ = 3 μm, $q \parallel k_t$) in dependence on the angle of the incidence θ [274]. At a normal incidence the measured value $\widetilde{A}$ = 2,9%. For θ = 12,5^O the maximum value of the absorptivity was attained: $\widetilde{A}_q$ =15%. For the comparison of the present theory with the experimental data [274] we use the optical constants of Al: n = 30,2, m = 98,4 [208]. According to the formula (3.194), (3.196), (3.197) the absorptivity $\widetilde{A}$ = 100% is reached for θ = 12,74^O and $4a_{p0}$ = 0,44 μm, when θ = 12,74^O and $4a_q$ = 3 μm, then the absorptivity, calculated with the help of the formula (3.196), is equal to $\widetilde{A}$ = 10,4%, which is in satisfactory agreement with the experimental value [274]. For more detailed comparison of analytical formula (3.196), (3.190) with the experiment see Ref. [275].

3.8. The comparison of the linear theory of the interferencial instabilities of the surface relief and the experimental results on ripples generation

As was shown in Sec. 3.3 − 3.5, the characteristics of all laser induced instabilities are determined by the value of dielectric permeability $\epsilon(\omega)$. Because of this the most simple interpretation assumes the case of metals, in which the value of $\epsilon(\omega)$ in the lessest degree is subjected to variation during the action of the laser pulse.

In the all experiments done on metals [144–146] it was obtained, that at not too big angles $\theta \leqslant \theta \lesssim 35^O$ the gratings are generated with the fronts perpendicular to the projection of the field vector E_i on the plane z = 0 ($q \parallel E_i$). In the case of s-polarization one grating is generated with period $d \approx \lambda/\cos\theta$ and in the case of p-polarization − two gratings with the periods $d_{\pm} \approx \dfrac{\lambda}{1 \pm \sin \theta}$ (see Fig. 26).

From the linear laser-induced instabilities theory (SAW − Sec. 3.3, CW − Sec.3.4 or IEI − Sec. 3.5) point of view to the dominant gratings corresponds the surface waves with the maximum growth rates γ_q. When $\epsilon' < -1$, the predictions of the theory of all three instabilities are confirmed by above mentioned experimental results (compare Fig. 29, 30, 31, 35).

The gratings with the same orientation are dominant also in the case of semiconductors, subjected to the laser action of s- or p-polarized beam. If the semiconductor is illuminated by the laser pulses with circular polarization, then from the CW or IEI instability theory it follows that after sufficient exposition time all the gratings with $\gamma_q > 0$ must grow to apprecient extent, their amplitudes in linear approximation being greater with greater values of γ_q. The experiment [180] corroborates this conclusion (compare Fig. 42 with Fig. 30, 31). The sharp difraction maxima at $\mathbf{q} \perp \mathbf{k}_t$ on Fig. 42 correspond to sharp peaks of γ_q at $\cos\varphi_s \approx \sin\theta$ on Fig. 30, the broad maxima at $\mathbf{q} \uparrow\uparrow \mathbf{k}_t$ and $\mathbf{q} \uparrow\downarrow \mathbf{k}_t$ on Fig. 42 correspond to the values γ_q at $\cos\varphi_a = \pm 1$ on Fig. 31. At last the additive small maxima on Fig. 42 correspond to the local maxima at $\cos\varphi_a = 0$ on Fig. 30 or local maxima on Fig. 31. The theoretical consideration of the surface

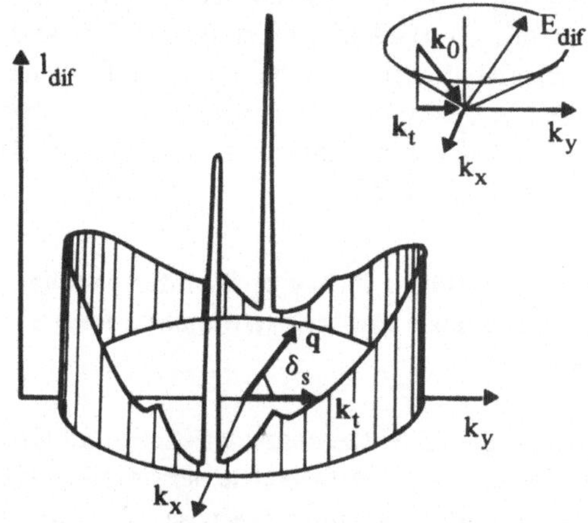

Fig. 42. Angular spectrum of diffraction of the probe wave, probing the complicated periodic structure frozen on the Ge surface. The structure was created by pulsed circularly polarized radiation with $\lambda = 1{,}06\ \mu m$ (the photograph of this surface is shown in Fig. 24c; from Ref. [180]). The positions of the peaks in the angular spectrum are in good agreement with the theory (Figs. 30 and 31).

pulse laser heating (see, below) shows that in typical PLA and surface ripples generation conditions in metalls and semiconductors the surface melting is attained. In [153, 202] the uniform melts presense was coroborated by a) appearance of high reflectivity phase; b) appearance of oscillations of probe beam diffraction intensity owing to the CW exitation. The fact that the characteristic intensity values, at which surface ripples are generated, are approximately one order of magnitude greater in metals then in semiconductors is then explaned by more high melting threshold in metals.

One must note, that at present there is no experiment which reveals uniquacally the feedback mechanism operating in surface ripples generations

152

process. For example in experiments [152, 153, 202] the pump pulse duration is of the same order of magnitude as the inverse frequency ω_0 of free CW. Thus in these experiments actually was recorded the fact, that surface relief perturbation arising during the laser pulse action, relaxes through exitation of decaying free CW. Thus it is expeditient to made the theoretical estimates, showing the sufficient effectivness of CW generation mechanism at pump intensity values which are characteristical to the experiments. Let us estimate the threshold light intensity needed for melting. From the solution of the non-stationary equation of the type of (3.62) at $v_0 = 0$, in zero approximation in $\xi_q(v \sim \xi_q)$, one gets the formula for the temperature (valid at times $\gamma_0^2 \chi t \gg 1$, $t \gg \dfrac{z^2}{4\chi}$)

$$T_0(z, t) - T_{in} \cong \frac{2f_0}{\gamma_0} \sqrt{\frac{t}{\pi \chi}}$$

From this equation, using for $f_0 - $ (3.60) in the case of s-polarization and (3.61) for p-polarization, one has for the threshold melting intensity $E_{melt\,a}^2(a{=}s,p)$ for metals ($\epsilon' \ll -1$, $|\epsilon'| \gg \epsilon''$)

$$|E|^2_{melt,p} = \frac{\pi \Delta T(\pi \kappa c_v)^{1/2}}{2c\tau_p^{1/2}} \frac{|\epsilon'|^{3/2}}{\epsilon''} \, , \quad |E|^2_{melt,s} = \frac{|E|^2_{melt,p}}{\cos^2\theta} \tag{3.202}$$

where $\Delta T = T_{melt} - T_{in}$.
Analogously for the dielectrics ($\epsilon' \gg \epsilon''$, $|\epsilon| \gg 1$) one obtains

$$|E|^2_{melt,p} = \frac{\pi \Delta T(\pi \kappa c_v)^{1/2}}{4c\tau_p^{1/2}} (\epsilon')^{1/2}, \quad |E|^2_{melt,s} = \frac{|E|^2_{melt,p}}{\cos^2\theta} \tag{3.203}$$

Let us make the numerical estimates, which show the possibility of effective CW — generation in Ge in experimental conditions [153, 202] (where $\lambda = 1,06\mu m$, $\tau_p = 20$ ns, $I_i = 2,5 \cdot 10^7$ W/cm^2). The melting threshold according to (3.203) is $I_{melt} = 6,4 \cdot 10^6$ W/cm^2 (for solid Ge; $\rho = 5,46$ g/cm^3, $c_v(400\,K) = 1,87 \cdot 10^7$ erg/cm^3, χ (275 K) $= 0,6 \cdot 10^7$ erg/cm s degree, $T_{melt} = 1210$ K [182], $\epsilon = 16 + i0,8$ [153]. The surface statical grating exitation threshold in the melt is in accordance to (3.91) $I_{th} = 6,1 \cdot 10^6$ W/cm^2 ($\theta = 0$, $q = k_0 = 2\pi/\lambda$, $\Omega_q = 0$ (the data for liquid Ge needed for calculation are given in Fig. 30 caption).

As $I_{th} < I_{melt}$, the statical gratings generation is possible just after surface melting. At intensity I_{th} =9,3 $\cdot 10^6$ W/cm^2, according to (3.203) the melting is attained in 9,5 ns, during remaining $\Delta\tau$ = 10,5 ns according to (3.90), (3.91) the grow of static grating with q $\parallel$ E_i, q $\cong$ k_0, growth rate γ_q = 10^8 c^{-1} takes place, so that $\gamma_q\Delta\tau$ = 1. Thus, at all values $I_i > I_{th}$ (in particular at intensities used in [153, 202]), during the time τ_p the amplitude of statical grating will grow to appreciate value. After the end of the laser pulse the surface perturbation leads to the exitation of two free CW (Ω_q = ω_0) running in opposite directions and decaying with decay constant γ_q = $2\nu q^2$ ~ 10^7 s^{-1}, which was recorded in [153, 202].

Analogous estimate can be made for Al. In experiment [145] λ = 1,06 μm, τ_p = 7 ns, I_i − 4 $\cdot 10^8$ W/cm^2. The data needed for calculations are taken from [181, 182, 208]. At this pump intensity, according to (3.202), (3.90), (3.91) the Al surface is melted in 4 ns. Then during the left $\Delta\tau$ = 3 ns the grow of grating with q $\cong$ k_0, q $\parallel E_i$ and growth rate γ_q =10^9 s^{-1} occurs. We thus have $\gamma_q\Delta\tau$ = 3, which shows the possibility of appriciable grating amplitude grow due to the CW exitation mechanism. The theory of laser induced CW instability (Sec. 3.4) explains also the experimental fact of changing the dominant gratings vector q orientation under the grow of the angle θ in the region $\theta \gtrsim 35^o$; for example at θ = 60° the angle δ_s (see Fig. 27b), according to data [139] is δ_s = 46°. As one can see from Fig. 30, the absolute maximum value for $0 \leqslant \theta \leqslant 30^o$ when $\cos\varphi_s$ = $n^{*-1}\sin\theta \approx \sin\theta$ (i.e. q $\perp E_{it}$), and when θ increases from 30° to 40° the absolute maximum of γ_q is attained at $\cos\varphi_s$ = 0. At θ = 60° the theoretical value of δ_s, according to (3.95) is δ_s = 49°, which is near to the above mentioned experimental value. As have been already noted, in the frame of CW instability theory the more subtle effects are explained: the "double" structure generation − two gratings with almost the same orientations (Sec. 3.4.5) and small scale structures (Sec. 3.4.6).

One can thus draw the conclusion that the mechanism of CW instability is responsible, in the main, for the generation of irreversible gratings on the surfaces of metals and semiconductors (the IEI and recoil pressure mechanism may give the additive contributions, see Sec. 3.4 and [169]). At the same time the appearance of the reversible gratings, existing only during pump pulse, may be explained by SAW generation in regimes without melting. Thus in experiments [150] at the intensities I_i = 5 $\cdot 10^7$ − 2 $\cdot 10^8$ W/cm^2, λ = 1,06 μm, τ_p = 200 ns, the probe beam diffraction has registered the reversible changes, and at $I_i > 2 \cdot 10^8$ W/cm^2 the changes became irreversible. According to the estimate of Sec. 3.3 the critical intensity of SAW generation at these conditions is I_{th} = 2 $\cdot 10^7$ W/cm^2. The theoretical estimate of melting threshold for Cu by the equation (3.202) is 2,8 $\cdot 10^8$ W/cm^2 (in solid Cu ρ=8,93 g/cm^3,   (573)=

$= 3,73 \cdot 10^7$ erg/cm^{2}s deg, $c_v(600K) = 3,71 \cdot 10^7$ erg/cm^{2}s deg, $T_{melt} = 1357$ K [182], $\epsilon = -37 + i3,4$ ($\lambda = 1$ μm) [208]). Both theoretical estimates are in a good agreement with experiment [150]. Let us now consider the peculiarities of gratings generation in dielectrics (crystalline and fuzed quartz [149]) under the action of CO_2 laser radiation. At $\lambda = 10,6$ μm ($\nu \equiv 1/\lambda = 943,4$ cm^{-1}) or at $\lambda = 10,27$ μm ($\nu = 973,7$ cm^{-1}) the gratings with usual orientation ($q \parallel E$) arise on the surface, one- at s-polarization and two gratings at p-polarization (see Fig. 26). When, however, the radiation with $\lambda = 9,33$ μm ($\nu = 1072$ cm^{-1}) was used for the surface exitation, the situation radically changed. At all $\theta \geqslant 0$ the gratings with transverse orientation $q \perp E$ were generated (two in the case of s-polarization, and one for p-polarization). As one can see from work [149] at $\theta = 0$ the periods of longitudinal gratings $d < \lambda$, and the periods of transverse ones $d > \lambda$, both these experimental results correspond to IEI theory predictions (see Table V). According to the results of IEI theory [172] (Sec. 3.5) transverse gratings must be generated under the condition $|\epsilon'| < 1$, $|\epsilon'| \gg \epsilon''$ (the same conclusion is valid in the cases of SAW or CW instabilities under the condition $|\epsilon'| < 1$, $|\epsilon'| \gg \epsilon''$, which are not considered here explicitly). In the work [149] the value of ϵ' in polariton region of quartz glass is estimated. In the temperature region of quartz glass softening $T = 1100°C$ the value $\epsilon'(1040$ cm$^{-1}) = 0$, and $\epsilon'(1120$ cm$^{-1}) = -1$. From this estimate one can see that for $\lambda = 9,33$ μm inequality $|\epsilon'| < 1$ holds. From the standpoint of IEI theory (and also SAW or CW generation) (see Table V) at $\lambda = 9,33$ μm the transverse gratings must be generated, which is corroborated by experiment. The comprehensive experimental and theoretical study of surface ripples generation in fused silica was performed in the work [267]. The link between the surface ripples types and value of ϵ was also pointed out in the works [166, 143].

All the above considered examples refer to the case of laser wave incident from vacuum into the medium with dielectric permeability ϵ. In more general case of wave incidence from the medium with dielectric permeability ϵ_1 into the medium with dielectric permeability ϵ the resonance denominator in equations (3.8) − (3.10) is replaced by symmetrical expession $D_p = \epsilon(k_a^2 - \epsilon_1 k_0^2)^{1/2} + \epsilon_1(k_q^2 - \epsilon k_0^2)^{1/2}$. Under the condition that $|\epsilon| \gg |\epsilon_1|$, $|\epsilon| \gg 1$, we have $D_p \sim (k_q^2 - \epsilon_1 k_0^2)^{1/2} - \epsilon_1 k_0/(-\epsilon)^{1/2}$, that is the sharp resonance appears at $k_a \approx (\epsilon_1)^{1/2} k_0$. In this case one may expect that the ripples period would be $\dfrac{\lambda}{\epsilon_1^{1/2}}$. We note, the ripples with this period were observed on the surface of wide gap dielectrics in experiments [203], where the laser radiation was focused on the output face of plane-parallel plate. The required inequality $|\epsilon| \gg |\epsilon_1|$, may be fulfilled due to the melting or plasma formation at the output face.

Thus, one may draw a general conclusion, that predictions of linear laser-induced instabilities theories (SAW, CW-exitation or IEI) are, in the main, in a good agreement with the experiments and that the character of the gratings, formed on the surface is determined by the value of dielectric permeability of the medium $\epsilon(\omega)$ during the SPS formation.

4. THE DIFFUSIINAL-DEFORMATIONAL INSTABILITIES AND PHASE-TRANSITIONS ON THE SURFACE UNDER THE ACTION OF LASER RADIATION

The instabilities, considered in Sec. 3 may be called the interferencial instabilities (II). The SPS, which are formed owing to the II are characterized by the dependence of their orientations relative to the incident wave electric field vector E_i orientation and their periods $d \sim \lambda$.

In this section we shall consider another broad class of surface laser-induced instabilities discovered and investigated recently, which are qualitativly different from II — the diffusional-deformational instabilities (DDI). DDI can take place in a wide class of the systems, in which some diffusional (or kinetical) variable (the nonequibrium carriers concentration or temperature in the semiconductors, the concentrations of the vacancies, dislocations, voids or the concentration of the amorphous phase in the crystalising medium) is, in the presence of laser field or, in general, of some nonequilibrium exitation, coupled with the deformation of the medium. The initial deformation spatially (and temporally) modulates the activation energy of the diffusional variable W and thus its concentration, which leads to the appearance of the forces, proporional to the gradients of the concentrations. These forces enhance the initial deformation and lead to the development of DDI, when the intensity of laser field exceeds the critical value I_{th}. The geometry of the SPS formed as a result of DDI does not depend on the polarization of laser field and the periods of SPS are not determined directly by the value of λ. The symmetry of these SPS is determined either by the symmetry of the medium or by that of the laser field.

Below we consiider the examples of the endothermal DDI, which develop owing to the laser pulse energy and also of the exothermal DDI, which develop owing to the latent heat of the medium with laser radiation playing the role of the initial activation. We shall consider also the experimental results, which can be interpreted from the viewpoint of DDI.

The prototype of DDI is the electron-deformational-thermal instability in semiconductors, which we consider in greater details.

4.1. The electron-deformational-thermal instability and the semiconductor-metal phase transition with the formation of the ordered surface structures

As was discussed in Sec. 1.3 the problems of pulsed laser modification of the surface layers of semiconductors (PLA [5—12] amorphisation [220, 276—278], semiconductor-metal phase transitions [70, 279, 280]) are actively studied at present. These phase transformations occur owing to the change of the state of the lattice, while laser radiation interacts primary with the electronic subsystem. The electron-phonon interaction and corresponding transformation of the absorbed energy, thus, plays the key role in the effects, mentioned above.

As it is was pointed out in Sec. 1 the traditional mechanism of energy relaxation in semiconductors, exited by pulsed laser radiation, consists in subsequent electron-electron, electron-phonon and phonon-phonon relaxation on times of order of picoseconds. The resulting spatial picture of the lattice modifications must then reflect the monotonous decrease of the radiation intensity with the distance from the centre of laser spot and also the monotonous decrease of the absorbed energy into the medium with a characteristic length equal or exceeding optical linear absorption length γ_0^{-1}.

A number of recent experimental results, however, obtained with the help of powerfull ultrashort laser pulses, cannot be explained on the basis of these traditional ideas. The formation of ordered surface structures is often observed instead of monotonous picture. For example, the formation of the spatially periodical structures with the alternation of the metalic and dielectric phases is observed under the irradiation of the semiconductor phase of VO_2 by picosecond laser pulses [280]. The symmetry of these structures is determined by the symmetry of the surface or that of laser field: the concentric rings in the center and the radial rays in the outside region of the irradiated spot ("the sun") are formed for the axial symmetry of laser field and the one dimensional gratings − for the one dimensional symmetry of the field (see Fig. 43). Pulsed laser irradiation of the surface of semiconductors leads also to the appearance of the concentric ring structures, formed by the alternation of the crystalline and amorphous phases [281]. A number of the experimental results point out on the nonlinear mechanism of the laser energy transfer to the lattice at the strong pumping intensities. So, in Ref. [282] the order of magnitude decrease of the absorption length relative to γ_0^{-1} was recorded. In Ref. [126, 292] it was found that the powerfull femtosecond pulses lead to the melting of the surface of the semiconductor over the times of the order of their duration much shorter then τ_{e-T}.

The experimental facts, mentioned above, may be interpreted from the viewpoint of some laser-induced instability, leading to the periodical (along the surface) change of the state of the lattice and the sharp increase of the effective absorption coefficient.

In this Section two new laser-induced instabilities developing on the surfaces of semiconductors are considered: the electron-deformational instability (EDI) and deformational-thermal instability (DTI), the theory of which can explain

158

the formation of the above mentioned ordered structures. These two instabilities can occur together, leading to the developing of the electron-deformational-
thermal instability (EDTI).

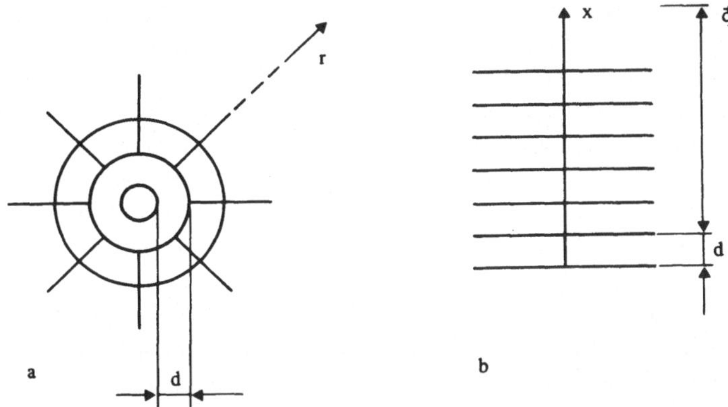

Fig. 43. The spatially-periodical pictures of the coupled surface fields of the deformation, the nonequilibrium carriers concentration and the temperature,
arising as a result of EDTI. a) the "sun" – like structure, the axial symmetry of the laser field and the isotropic surface; b) the grating (the direction of vector **q** is selected by one dimensional geometry of laser field
or by the crystallographic axis.

The physical mechanism of EDTI consists in the following. The surface deforma-
tion $\xi = \mathrm{div}\ \mathbf{u}$ (**u** is the displacement vector of the medium) and also the density
of the nonequilibrium carriers n and the temperature T spatially modulate
the forbidden gap width

$$E_g = E_{g0} + \theta\xi - \beta_n n - \beta_T T , \qquad (4.1)$$

where E_{g0} – the starting gap width, $\frac{\partial E_g}{\partial \xi} = \theta, \theta = \theta_{cc} - \theta_{vv}, \theta_{cc}, \theta_{vv}$ – the deformational potential for the conductance and valence bands correspondingly.
The phenomenological coefficient $\beta_n = \left| \frac{\partial E_g}{\partial n} \right| > 0$ takes into account the de-
crease of E_g owing to breaking of the covalent bonds, when the electrons are
exited from the bonding states in the valence band to the anti-bonding ones
in the conductance band [70], the coefficient $\beta_T = \left| \frac{\partial E_g}{\partial T} \right| > 0$ takes into
account the decrease of E_g owing to the heating [279]. The modulation of E_g
leads to the modulation of the interband absorption coefficient γ and, consequently to the further modulation of n and T. This leads to the appearance

159

of the forces $\mathbf{F}_n = \theta_{grad}\,n$ and $\mathbf{F}_T = -K_a \mathrm{grad}T$, which ampliphy the starting displacements of the lattice, leading to the instability of the amplitudes of the surface deformations, the density n and of the temperature T, when the pumping intensity exceeds certain critical value. The consideration of the EDTI-mechanism was carried out in Ref. [190, 283, 307].

In this Section the boundary problem of EDTI development on the surface of the semiconductors under the condition of a strong optical absorption is solved. The dispersion equation for EDTI, which determines the dependence of the growth rate $\gamma_q \equiv \lambda_q$ and the frequency $\lambda''(q)$ of the Fourie-amplitudes of the coupled fields n, T, ξ on the wave vector $\mathbf{q}$, $(\lambda = \lambda(q) = \lambda' + i\lambda'')$ is obtained. From it follows, that under EDTI three qualatatively different instabilities can develop.

The first one is the instability of the surface acoustical waves (SAW) $(\lambda'' \neq 0, \lambda' > 0)$, which develops, starting from the thermal fluctuating SAW. The second is the softening of the frequencies of SAW $(\lambda'' \to 0, \lambda' < 0)$. Finally, the third one is the instability of the static deformations $(\lambda'' = 0, \lambda' > 0)$, as a seeds for which serve the quasistatic surface deformations arising owing to the presence of the fluctuations of the temperature (or density of nonequilibrium carriers). The detailed study of latter istability is carried out in this Section. It is shown, that the EDTI development leads to the formation of the complicated ordered configurations of the coupled surface fields ξ, n and T (in the form of the gratings, rings, rays, "suns" and radial-ring cells). The periods of these structures in dependence on the parameters of the medium and the intensity and the duration of laser action are determined, as well as the depth of their penetration from the surface into the medium.

The stationary nonlinear regime of EDTI is considered, the stabilization of which occurs owing to the nonlinear Auger-recombination of the carriers (and, possible, to the optoacoustical nonlinearity). The stationary Fourier-amplitudes of the deformation, the carrier density and the temperature in dependence on the wave vector $\mathbf{q}$, and the pumping intensity are found. It is shown, that the amplitude of the modulation (along the surface) of the gap width, according to (4.1), can lead to the spatially periodical cancellation of E_g, that is to the laser-induced semiconductor-metal phase transition, accompanied by the surface ordered structures formation.

The dielectric-metal phase transition under the action of light with the cancellation of E_g owing to the terms β_n and β_T in the formula (4.1) was studied previously in Ref. [70, 279]. The account of the deformation and the effect of the surface made in this work are the new contribution into the theory of laser-induced semiconductor-metal phase transition, which enables to explain the phonomenon of the formation of the ordered structures on the surface

160

observed in Ref. [280]. The estimates of the parameters of these structures shows that the EDTI mechanism may be responsible for their formation.

4.1.1. The closed system of equations for the modulations of the density of the nonequilibrium carriers, the temperature and the vector of the displacement of the medium

Let the semiconductor occupying the half space $z > 0$ is irradiated by the laser wave

$$\mathbf{E}(\mathbf{r}, t) = \mathbf{E} \exp(-i\omega t + ikz) + c.c., \tag{4.2}$$

where $\mathbf{E} = \mathbf{E}(\mathbf{r}) f(t)$, $f(t) = 1$ for $0 \leqslant t \leqslant \tau_p$ and $f = 0$ for $t < 0, t > \tau_p$.
In the case considered ($\epsilon' \gg 1$, $\epsilon' \gg \epsilon''$) the equations for the concentration of nonequilibrium carriers n and temperature of the medium T have the form

$$\frac{\partial T}{\partial t} - \chi \Delta T = 2\omega \epsilon'' |E|^2 e^{-\gamma z} / \pi c_v \epsilon', \tag{4.3a}$$

$$\frac{\partial n}{\partial t} - D\Delta n + \frac{n}{\tau} + \gamma_A n^3 = 2\epsilon'' |E|^2 e^{-\gamma z} / \pi \hbar \epsilon', \tag{4.3b}$$

where $\gamma = \omega \epsilon'' / c\sqrt{\epsilon'}$ is the optical absorption coefficient, D is the diffusion coefficient of the carriers, τ is the time of linear recombination, γ_A is the constant of Auger recombination.
Due to (4.1) the optical constants ϵ'' and ϵ' depend on n, T and ξ. In the range of the values of n, T and ξ which are of interest here, one can represent this dependency for ϵ'', using the formula for the interband dielectric permeability of the semiconductor, with taking into account of (4.1), in the form

$$\epsilon'' = \epsilon_0'' + \frac{\partial \epsilon''}{\partial \omega} \frac{1}{\hbar} (\beta_n n + \beta_T T - \theta \xi). \tag{4.3c}$$

Due to (4.4), as it is seen from (4.3), the diffusional fields n and T are coupled (in the presence of laser field) to the deformational field ξ.
Let us represent the solutions of (4.3) in the form
$$n = n_0 + n_1, \quad T = T_0 + T_1, \quad \xi = \xi_0 + \xi_1, \quad U = U_0 + U_1,$$
where n_0, T_0, ξ_0, U_0 are the solutions of (4.3a,b) with taking into account formula (4.3c), taken in zeroth approximation in n, T, ξ. The explicit form of these of these solutions is not important here. Let us linearize the equations

(4.3a,b) with taking account of (4.3c). Then we obtain the linear equations for the density of the nonequilibrium carriers n_1 and the temperature T_1 which can be written in the unified form

$$\frac{\partial Y_{j1}}{\partial t} + \Delta_j Y_{j1} = (\epsilon_{j\xi}\xi + \epsilon_{jT}T_1 + \epsilon_{jn}n_1)\, e^{-\gamma_0 z}(1 - \gamma_0 z)\,, \qquad (4.4)$$

where $j = n, T$, $Y_n = n_1$, $Y_T = T_1$, $\Delta_j = -\chi_j \Delta + \tau_j^{-1}$ is the operator which for $j=n$ takes into account the diffusion and the recombination of the carriers, $\chi_T \equiv \chi$, $\chi_n \equiv D$, $\tau_T^{-1} = 0$, $\tau_n = \tau^{-1} + 3\gamma_A n_0^2$. The coupling coefficients are determined by the formula

$$\epsilon_{ja} = 2\frac{\partial \epsilon_0''}{\partial \omega}|E|^2\, \sigma_{ja}/\epsilon'\pi\hbar\,, \qquad a = n, T, \xi \qquad (4.5)$$

$$\sigma_{n\xi} = -\theta/\hbar\,, \quad \sigma_{nT} = \beta_T/\hbar\,, \quad \sigma_{nn} = \beta_n/\hbar\,, \quad \sigma_{T\xi} = -\omega\theta/c_v\,,$$

$$\sigma_{TT} = \omega\beta_T/c_v\,, \qquad \sigma_{Tn} = \omega\beta_n/c_v\,.$$

The boundary conditions for χ_{j1} have the form

$$\frac{\partial Y_{j1}}{\partial z}\bigg|_{z=0} = 0\,, \qquad Y_{j1}\big|_{z\to\infty} = 0\,. \qquad (4.6)$$

The system (4.4–4.6) is closed with the equation for the vector of the displacement of the medium **U**. In the case of isotropic medium it has the form

$$\frac{\partial^2 U}{\partial t^2} = c_t^2\,\Delta U + (c_l^2 - c_t^2)\,\mathrm{grad}\,\mathrm{div}\,U + \sum_{j=0,T} f_j\,\mathrm{grad}\,Y_j\,, \qquad (4.7)$$

where $f_n = \theta/\rho$, $f_T = -K\alpha/\rho$. The form of the boundary conditions for **U** depends on the summetry of laser field or that of the surface (see Sec. 4.1.2). The system of equations (4.4)–(4.7) can be solved analitically in two limiting cases: a) a weak optical absorption in the bulk and; b) a strong optical absorption, when γ_0^{-1} is smaller than the depths of penetration into the medium of the surface material exitations n_1, T_1, ξ_1 (the surface case). Here we shall confine our consideration by the latter case, which is more interesting from the practical viewpoint. Then the equation (4.4) can be written in a more simple form:

$$\frac{\partial Y_{j1}}{\partial t} + \Delta_{j0} Y_{j1} = (\epsilon_{j\xi}\xi + \epsilon_{jT}T_1 + \epsilon_{jn}n_1)\bigg|_{z=0} e^{-\gamma_0 z}\,, \qquad (4.8)$$

where $\quad \Delta_{n0} = -D\Delta + \tau_{n0}^{-1}$, $\quad \tau_{n0}^{-1} = \tau^{-1} + 3\gamma_A \, n_0^2\,(0)$, $\quad n_0\,(0) = n_0\big|_{z=0}$.

4.1.2. The general dispersion equation of the electron-deformational-thermal instability. The one dimensional gratings formation.

Let, owing to the symmetry violation of the laser field, or that of the medium, some specific direction x is selected. Then the boundary conditions at the surface $z = 0$ for the vector $\mathbf{U} \equiv \mathbf{U}_1$ have the form

$$\frac{\partial U_x}{\partial z} + \frac{\partial U_z}{\partial x} = 0 \; ; \quad \sum_{j=n,T} \frac{f_j\,Y_{j1}}{c_l^2} + \frac{\partial U_z}{\partial z} + (1-2\beta)\,\frac{\partial U_x}{\partial x} = 0 \, , \qquad (4.9)$$

where $\beta = c_l^2/c_t^2$. Let us study the solution of the system of equations (4.6–4.9), putting in the formula (4.2) $E(\mathbf{r}) = \text{const}$.

Let us define the deformation ξ and the diffusional field Y_{j1} for $z = 0$ in the form

$$\xi = A \, \exp(iqx + \lambda t) \, , \qquad Y_{j1} = A_j \, \exp(iqx + \lambda t) \, , \qquad (4.10)$$

where $A = A(q)$ is the initial amplitude of the deformation, A_j — some constants. This form of solution defines the surface field of the deformation, density of the carriers and of the temperature in the form of the one dimensional gratings (see Fig. 43). We seek the solution of the equation (4.8), taking into account (4.10), in the form

$$Y_{j1} = (B_j \, e^{-\gamma_0 z} + c_j \, e^{-\delta_j z})\,\exp(iqx + \lambda t) \, . \qquad (4.11)$$

Substituting (4.11) into (4.8) and taking into account the boundary conditions (4.6) as well as the conditions of selfconsistency of the expressions (4.8) and (4.10), we find for $\gamma_0 \gg \delta_j$:

$$Y_{j1} = A\epsilon_{j\xi}\exp(iqx + \lambda t - \delta_j z)\,[\chi_j\delta_j\gamma_0(1 - \sum_{j=n,T} \epsilon_{jj}/\chi_j\gamma_0\delta_j)]^{-1} \, , \qquad (4.12)$$

where

$$\delta_j = (q^2 + (\lambda + \tau_{j0}^{-1})/\chi_j)^{1/2} \qquad (4.13)$$

163

Now we consider the solution of the equations (4.7) and 4.9). We represent the vector $\mathbf{U}$ in the form $\mathbf{U} = \mathbf{U}_1 + \mathbf{U}_t$, with

$$\operatorname{div} \mathbf{U}_t = 0 ; \qquad \operatorname{rot} \mathbf{U}_1 = 0 . \qquad (4.14)$$

The equations for the vectors $\mathbf{U}_t$ and $\mathbf{U}_1$ are obtained from the equation (4.7)

$$\frac{\partial^2 \mathbf{U}_a}{\partial t^2} = c_a^2 \Delta \mathbf{U}_a + \delta_{a,1} \operatorname{grad} \sum_{j=n,T} f_j Y_{j1} , \qquad (4.15)$$

where $a = 1, t$, $\delta_{a,1} = 1$, if $a = 1$, $\delta_{al} = 0$, if $a = t$; the quantities Y_{j1} are given by the formulas (4.12). The solution of the system (4.15) is given by the formula

$$\mathbf{U}_a = (\mathbf{B}_a e^{-\kappa_a z} + \delta_{a,1} \sum_{j=n,T} \mathbf{D}_j e^{-\delta_j z}) \exp(iqx + \lambda t) . \qquad (4.16)$$

Substituting this expression into (4.15), taking into account the conditions (4.14), we express the projections B_{1x}, B_{1z} through some constant M, and the projections B_{tx}, B_{tz} — through the constant N. From the condition of selfconsistency of the solutions (4.16) and the formula for $\xi|_{z=0}$ (4.10) we express the constant A through the constant M:

$$A = \frac{\lambda^2 M}{c_1^2} \left[1 - \sum_{j=n,T} \frac{R_j c_1^2 (q^2 - \delta_j^2)}{\chi_j \gamma_0 \delta_j (\lambda^2 + c_1^2 (q^2 - \delta_j^2))} (1 - \sum_{j=n,T} \epsilon_{jj} / \chi_j \gamma_0 \delta_j)^{-1} \right]^{-1} ,$$

$$(4.17)$$

where

$$R_n = \theta^2 \frac{\partial \epsilon''}{\partial \omega} |E|^2 / 2\pi \hbar^2 c_1^2 \rho ; \qquad R_T = -Ka \frac{\partial \epsilon''}{\partial \omega} \theta \omega E^2 / 2\pi \hbar c_v \rho c_1^2 .$$

The final solution is expressed through M and N and has the form

$$U_x = (\kappa_t N e^{-\kappa_t z} + iqM e^{-\kappa_1 z} - \lambda^2 \sum_{j=n,T} \frac{iq}{\delta_j} M \Phi_j e^{-\delta_j z}) e^{iqx + \lambda t} ,$$

$$U_z = (iqN e^{-\kappa_t z} - \kappa_1 M e^{-\kappa_1 z} + \lambda^2 M \sum_{j=n,T} \Phi_j e^{-\delta_j z}) e^{iqx + \lambda t} . \qquad (4.18)$$

164

Here

$$\kappa_{t,l} = (q^2 + \lambda^2/c_{t,l}^2)^{1/2} \, , \tag{4.19}$$

$$\Phi_j = R_j \, [\chi_j \, \gamma_0 \, (\lambda^2 + c_l^2 (q^2 - \delta_j^2))]^{-1} \left[1 - \sum_{j=n,T} \left(\frac{\epsilon_{jj}}{\chi_j \gamma_0 \delta_j} + \frac{R_j c_l^2 (q^2 - \delta_j^2)}{\chi_j \gamma_0 \delta_j (\lambda^2 + c_l^2 (q^2 - \delta_j^2))} \right) \right]^{-1} \tag{4.20}$$

Substituting the formulas (4.18) into (4.9) we obtain two homogenious equations for M and N:

$$N(\kappa_t^2 + q^2) + 2iqM(\kappa - \lambda^2 \sum_{j=n,T} \Phi_j) = 0 \, ,$$

$$-2i\kappa_t qN + (\kappa_t^2 + q^2) M (1 - \lambda^2 \sum_{j=n,T} \Phi_j/\delta_j) = 0. \tag{4.21}$$

Equating the determinant of the system (4.21) to zero, we obtain the general dispersion equation of EDTI (it holds also for the case of the structures with axial symmetry, see Sec. 4.1.4)

$$4q^2 \kappa_t \kappa_l - (\kappa_t^2 + q^2)^2 =$$

$$= \frac{\lambda^2}{c_l^2} \sum_{j=n,T} \frac{R_j (4q^2 \kappa_t \delta_j - (\kappa_t^2 + q^2)^2)}{\chi_j \gamma_0 \delta_j (\frac{\lambda^2}{c_l^2} + q^2 - \delta_j^2) - \sum_{j=n,T} (\frac{\lambda^2}{c_l^2} \epsilon_{jj} + (R_j + \epsilon_{jj})(q^2 - \delta_j^2))} \tag{4.22}$$

The dispersion equation (4.22) has three qualitativly different types of solution, which correspond to three types of EDTI. Let us consider each one seperately, confining itself in (4.22) with taking into account either EDI under the condition that $R_n/D\delta_n \gg R_T/\chi\delta_T$, or TDI under the fulfilment of the inverse inequality:

a) **Laser generation of SAW:** $\lambda = \lambda' + i\lambda''$, $\lambda' > 0$, $\lambda'' \gg \lambda$.

The dispersion equation (4.22) was obtained with neglect of the viscosity of the medium. To take into account the viscosity in the equation (4.22) and also in the formula for Φ_j, $\kappa_{l,t}$, R_j we change $c_{l,t}^2$ for $c_{l,t}^2(1 + \eta_{l,t}/\rho c_{l,t}^2)$, where $\eta_t = \eta$, $\eta_l = 4/3\,\eta + \zeta$; η, ζ are the first and the second coefficients of viscosity. Then, separating in the equation (4.22) the real and imaginary parts under the conditions $(\lambda + \tau_{j0})^{-1}/\chi_j \ll \lambda^2/c_{t,l}^2$, q^2, $2\lambda\chi_j/c_l^2 \ll 1$, $qR_j\chi_j/\chi c_l^2$, we obtain for the frequency of SAW the usual expression:

$\lambda'' = \sigma c_t q$, $0,87 < \sigma < 0,95$ [155]. From the imaginary part of (4.22) we obtain the expression for the growth rate

$$\lambda'' \cong -\frac{\sigma^2 \eta q^2}{2\rho} - \sum_{j \neq n,T} \frac{R_j \beta q}{\gamma_0} \tag{4.23}$$

In the derivation of (4.23) we used the conditions $\lambda''/D^2 q^4 \ll 1$ for EDI or $\lambda''/\chi^2 q^4 \ll 1$ for DTI. From (4.18) and (4.23) it is seen that SAW generation ($\lambda' > 0$) owing to EDI is possible only in the case $\partial\epsilon''/\partial\omega < 0$, and owing to DTI — in the case $\theta\,\partial\epsilon''/\partial\omega > 0$.

From (4.23) one obtains the critical intensity

$$|R_j|_c = \eta\gamma_0 q \sigma^2 / 2\beta\rho . \tag{4.24}$$

b) The softening of the acoustical frequencies: $\lambda' < 0$, $\lambda'' \to 0$.

Separating in (4.22) the real and the imaginary parts under the conditions λ, $\lambda'' < qc_{t,l}$, we obtain the expression for the acoustical frequency, renormalized by the pumping wave

$$\lambda''^2 = \Omega_q^2 \left(1 - 2\frac{\beta}{1-\beta} \frac{qR_j}{(q+\sqrt{q^2+\tau_{j0}^{-1}/\chi_j})(\chi_j\gamma_0\sqrt{q^2+\tau_{j0}^{-1}/\chi_j}-R_j-\epsilon_{jj})}\right)$$

$$\Omega_q = 2(1-\beta)^{1/2} qc_t/(3-2\beta)^{1/2} \tag{4.25}$$

As one can see from (4.25) and (4.17) the softening of the acoustical frequencies occurs in the case of EDI for $\partial\epsilon''/\partial\omega > 0$ and in the case of DTI for $\theta\,\partial\epsilon''/\partial\omega < 0$. The critical intensity is determined from the condition $\lambda'' = 0$. In fact, the decrease of the acoustical frequencies is limited by the values $\lambda' \sim \beta^2 q^2 \eta \ll \Omega_q$.

c) Laser generation of the static surface structures.

Putting in (4.22) $\lambda'' = 0$ and defining $\lambda' \equiv \lambda$, we obtain, under the condition $\lambda^2/c_{t,l}^2 < q^2$, $(\lambda + \tau_{j0}^{-1})/\chi_j$ the following equation

$$(q + \delta_j)(\delta_j - q_{R\epsilon}) = \frac{2\beta}{1-\beta} qq_R , \tag{4.26}$$

where
$$q_R = R_j/\chi_j\gamma_0 , \qquad q_{R\epsilon} = (R_j + \epsilon_{jj})/\chi_j\gamma_0$$

166

The solution of (4.26) has the form

$$\lambda = \chi_j \left[\left(\frac{(q+q_{R\epsilon})^2}{4} + q q_R \frac{2\beta}{1-\beta} \right)^{1/2} + \frac{q_{R\epsilon}-q}{2} \right]^2 - \chi_j q^2 - \tau_{j0}^{-1} \qquad (4.27)$$

The growth rate attains the maximum value λ_m in the point q_m:

$$q_m = q_{R\epsilon}[(5a^2 + 6a + 1)^{1/2} - (2a+1)], \qquad a = \frac{2\beta}{1-\beta}(q_R/q_{R\epsilon}) \qquad (4.28a)$$

$$\lambda_m = \chi_j q_{R\epsilon}^2 \frac{1}{2} [(5a+1)(5a^2 + 6a + 1)^{1/2} - 11a^2 - 8a + 1] \qquad (4.28b)$$

The dependence of λ on q is given in Fig. 44a for EDI and in Fig. 44b — for DTI. For $R_j \gg \epsilon_{jj}$ (the deformational limit) the maximum of $\lambda(q)$ selects the dominant structure with $q = q_m \neq 0$; for $R_j \ll \epsilon_{jj}$ (the diffusional limit) the maximum of λ is in the point $q = 0$. However, in this case as well the dominant structure with $q \neq 0$ is selected.

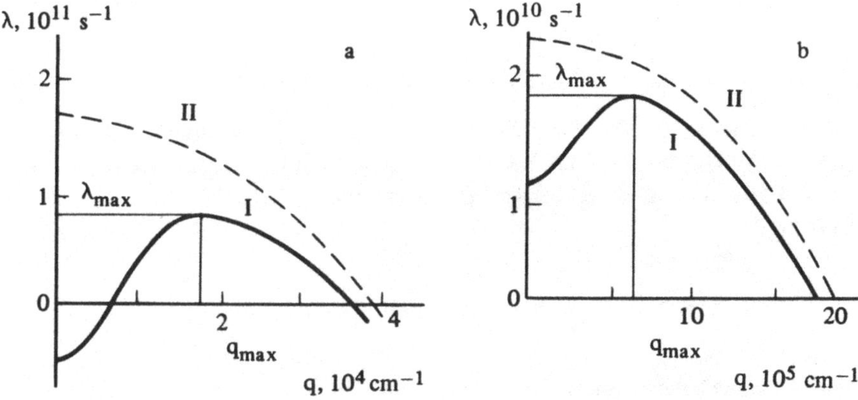

Fig. 44. The dependence of the growth rates λ of the statical SPS on the modulus of wave vector $\mathbf{q}$: a) EDI; b) DTI. The curves I corespond to the condition $R_j \gg \epsilon_{jj}$ (deformational limit) and are computed for the characteristic values of the semiconductor parameters, given in Sec. 4.1.5. The curves II correspond to the case $R_j \ll \epsilon_{jj}$ (diffusional limit) (schematically).

To show this, let us consider in greater details the nature of the initial fluctuations for EDTI. For t=0 the quantities $Y_{n1}(t{=}0) \equiv n_1(t{=}0)$ and $Y_{T1}(t{=}0) \equiv T_1(0)$ in formula (4.9) are the Fourier amplitudes of the initial fluctuations of the carrier concentration and the temperature on the surface (the correlator

$\langle Y_{j1}(t{=}0) \; Y_{j1}(t{=}0)\rangle$ = const and does not depend on q). Using the formula (4.18) and the connection $N = -2iq\kappa_1 M/(\kappa_t^2 + q^2)$, which follows from (4.21) taking for $t = 0\, R_j$, $\epsilon_{jj} = 0$ and also $\lambda = \lambda(t{=}0) = -\, \chi_j q^2 - \tau_{j0}^{-1}$, we obtain from (4.9)

$$M = -f_j \, Y_{j1}(t{=}0) \, / \, (1-\beta) \, (\chi_j q^2 + \tau_{j0}^{-1})^2 \; . \qquad (4.29)$$

The formulas (4.18) and (4.29) with taking into account the connection between M and N yield the final solution for the vector U under EDTI with the formation of ordered structures. In the diffusional limit ($\epsilon_{jj} \gg R_j$) the growth rate is determined by the expression, which follows from (4.27)

$$\lambda = \chi_j \, (q_{R\epsilon}^2 - q^2 - \tau_{j0}^{-1}/\chi_j) \; . \qquad (4.30)$$

Taking into account (4.29) and (4.17) we obtain the formula for the deformation amplitude on the surface, which occurs in (4.10) at t=0

$$A = A(q,t) = -\, \frac{\lambda^2}{\chi_j q^2 + \tau_{j0}^{-1}} \; \frac{f_1 \, Y_{j1}(t{=}0)}{\Phi c_1^2 \, (1-\beta)} \; ,$$

where λ is given by the formula (4.30) (for $\epsilon_{jj} \gg R_j$ the function Φ in (4.17) does not depend on q). With taking into account this expession, we write down from (4.10) the resulting deformation after the end of the pulse ($t = \tau_p$) in the form

$$\xi(x, \tau_p) = \sum_q A(q, \tau_p) \exp(iqx + \lambda\tau_p) \; .$$

Then the correlator of the deformations is

$$\langle \xi^2(x)\rangle = \sum_q \langle A_q^2(q,\tau_p)\rangle \, e^{2\lambda\tau_p} \; \sim \int_0^\infty dq \, \langle A^2(q,\tau_p)\rangle \, e^{2\lambda\tau_p} \sim \int_0^\infty \xi_q^2 \, dq \; ,$$

where the spectral function of the deformation is

$$\xi_q^2 \sim \left[\frac{q_{R\epsilon}^2 - q^2 - \tau_{j0}^{-1}/\chi_j}{q^2 + \tau_{j0}^{-1}/\chi_j} \right]^4 \, e^{2\lambda\tau_p} \; ,$$

168

and λ is given by the formula (4.30). The value $q = q_m$, for which the maximum of the spectral function is reached at $t = \tau_p$, determines, in the linear regime, the period of the grating in the case $\epsilon_{jj} \gg R_j$. For example, for the case of DTI $(\tau_{j0}^{-1} = 0)$ we have

$$q_m = q_{R\epsilon}, \qquad \text{for} \quad q_{R\epsilon} > (4\tau_p \chi)^{1/2},$$

$$q_m = (4\tau_p \chi)^{1/2}, \quad \text{for} \quad q_{R\epsilon} < (4\tau_p \chi)^{1/2}. \tag{4.31}$$

To obtain the expession for the critical intensity of EDTI ($\lambda = 0$), we expand (4.22) in the power series of $\lambda/c_t q$). Then we obtain the soft relaxation mode

$$\lambda = -(2\chi_j q^2 + \tau_{j0}^{-1}) + \frac{2q}{\gamma_0}\left(\frac{R_j}{1-\beta} + \epsilon_{jj}\right),$$

where R_j is given by (4.17), and ϵ_{jj} — by (4.5). In particular for EDI ($j = n$) we have the soft concentration mode

$$\lambda = -(2Dq^2 + \tau^{-1}) + \frac{2q}{\gamma_0}\,\frac{|E|^2}{2\pi\hbar^2}\,\frac{\partial\epsilon_0''}{\partial\omega}\left[\left|\frac{\partial E_g}{\partial n_0}\right| + \frac{\theta^2}{(1-\beta)\rho c_1^2}\right]. \tag{4.32}$$

It is evident from this expression, that EDI occurs only for $\partial\epsilon''/\partial\omega > 0$. This condition is met for different semiconductors in the spectral range of commonly used lasers.

For DTI ($j = T$) we have the soft temperature mode

$$\lambda = -2\chi^2 + \frac{2q}{\gamma_0}\,\frac{|E|^2\omega}{2\pi\hbar c_v}\,\frac{\partial\epsilon''}{\partial\omega}\left[\left|\frac{\partial E_g}{\partial T_0}\right| - \frac{\theta K a}{(1-\beta)\rho c_1^2}\right]. \tag{4.33}$$

It is seen, that DTI can arise for the different signs of $\partial\epsilon''/\partial\omega$, depending on the sign of θ and the relative magnitude of the first and the second terms in the brakets of (4.33). The condition $\lambda' = 0$, used in (4.32) and (4.33) determines the critical intensity E_c^2 of EDI and DTI, correspondingly.

Thus, from the consideration of this section it follows, that for $E^2 \geqslant E_c^2$ the certain Fourie-amplitudes of the coupled surface fields of the static deformation (4.18), of the density of the nonequilibrium carriers (4.12) and of the temperature (4.12) grow exponentially in time. As a result of EDTI the structures in the form of one dimensional gratings can form with the period

$$d = 2\pi/q_m \tag{4.34}$$

where, in the linear regime of EDTI in the case $R_j \gg \epsilon_{jj}$, the value of q is given by the formula (4.28a) and for $R_j \ll \epsilon_{jj}$ — by (4.31).

4.1.3. The nonlinear regime of the electron-deformational-thermal instability and semiconductor-metal phase transition

We consider the nonlinear regime for the case of EDI, the stabilization of which occurs owing to the nonlinear Auger-recombination. Let us define at z = 0: $n_1(0) = n_q e^{iqx}$, $\xi(0) = \xi_q e^{iqx}$, $T_1(0) = T_q e^{iqx}$, and $E_g(0) = E_{g0} + E_{g1} e^{iqx}$. In the stationary regime ($\lambda' = 0$) we obtain near the threshold ($E^2 \sim E_c^2$), with the substitution $\tau_n^{-1} = \tau^{-1} + \gamma_A |n_1|^2$ in (4.32):

$$n(0) \equiv |n_q| = \frac{1}{\gamma_A^{1/2}} \sqrt{\frac{2q}{\gamma_0}\left(\frac{R_n}{1-\beta} + \epsilon_{nn}\right) - 2D_q^2 - \tau^{-1}} \, . \qquad (4.35)$$

The threshold of the appearance of the stationary values of $n_q - E_c^2$ of course coincides with the threshold of EDI — (4.32). Far above the threshold ($E^2 \gg E_c^2$) from (4.27) we obtain analogously

$$n_q = \frac{1}{\gamma_A^{1/2}}\left[D\left[\left(\frac{(q+q_{R\epsilon})^2}{4} + q q_{R\epsilon}\frac{2\beta}{1-\beta}\right)^{1/2} + \frac{(q_{R\epsilon}-q)}{2}\right]^2 - Dq^2 - \tau^{-1}\right]^{1/2} . $$
$$(4.36)$$

Thus, the dependency n_q on q follows the dependency $\lambda = \lambda(q)$. From (4.10) and (4.12) we obtain for the amplitude of the wave of the stationary deformation on the surface

$$\xi_q = n_q (D\gamma_0 \delta_n - \epsilon_{nn})/\epsilon_{n\xi} , \qquad (4.37)$$

where $\delta_n = (q^2 + (\tau^{-1} + \gamma_A n_q^2/D))^{1/2}$ and for the amplitude of the stationary temperature wave $T_q = \epsilon_{T\xi} D \delta_n n_q/\epsilon_{n\xi}$. We substitute (4.37) and T_q into (4.1). Then for the wave of the renormalization of the gap at z = 0, taking into account the expression for ϵ_{nn} from (4.5), we obtain for $E^2 > E_c^2$

$$E_g(0) = E_{g0} - E_{g1}(q)\cos qx ; \quad E_{g1} = D\delta_n n_q \left[\frac{2\pi\hbar^2\gamma_0}{\dfrac{\partial\epsilon''}{\partial\omega}|E|^2} + \frac{\hbar\omega}{\chi q c_v}\frac{\partial E_g}{\partial n}\right] \qquad (4.38)$$

We note, that the term β_n in (4.1) is exactly compensated by the part of the term from the right hand side of (4.37), that is the effective renormalization of E_g in (4.38) occurs owing to the deformational and temperature waves only. Under the condition that the amplitude of the modulation in (4.38) $E_{g1} > E_{g0}$, the grating of the metallic and dielectric phases arises on the surface. The period of this grating for $R_j \gg \epsilon_{jj}$ is determined by the value of q_m (4.28). In the inverse limit $R_j \ll \epsilon_{jj}$ the maximum of n_q from (4.36) is attained at $q=0$. In this case in (4.38) one must carry out the summation over the surface modes

$$E_g(0) = E_{g0} - \sum_q E_{g1}(q) \cos qx \equiv E_{g0} - (E_{g1})_{eff} \, .$$

From (4.36) one has for $\epsilon_{jj} \gg R$, $n_q \sim \sqrt{q_{R\epsilon}^2 - q^2}$ and, with the help of (4.39), (4.38), where $\gamma_A n_q^2 \gg Dq^2$, retaining the second term in the brackets, one obtains

$$(E_{g1})_{eff} \sim \int_0^{q_{R\epsilon}} d_q \cos qx (q_{R\epsilon}^2 - q^2) \sim \sin q_{R\epsilon} x \, . \qquad (4.39)$$

Thus, for $\epsilon_{jj} > R_j$ the grating with the period $d = 2\pi/q_{R\epsilon}$ is formed in the non-linear regime.

Besides the instability of the Fourie-amplitudes with $q \neq 0$, the amplitudes of the deformation, the carrier density and the temperature with $q = 0$ grow in time, which give rise to the spatially uniform decrease of E_g, i.e. to the increase of the optical absorption coefficient. This can explain the experimental fact of the sharp decrease of the optical absorption length in the semiconductors under the high level of exitation by the light with the energy of quantum $\hbar\omega > E_g$ [282].

4.1.4. The formation of radial — ringed structures

Let now the laser field has the axial (relative to z-axis) symmetry. Then in the cilindrical system of the coordinate (the point $z = 0$ is the centre of the laser spot) the boundary conditions on the free surface $z = 0$ for the vector $\mathbf{U}$ have the form

$$\frac{\partial U_r}{\partial z} + \frac{\partial U_z}{\partial 2} = 0 \; ; \quad \frac{1}{r} \frac{\partial U_z}{\partial \varphi} + \frac{\partial U_\varphi}{\partial z} = 0 \, , \qquad (4.40)$$

$$\sum_{j=n,T} f_j Y_{j1} / c_1^2 + \frac{\partial U_z}{\partial z} + (1 - 2\beta) \; \frac{\partial U_r}{\partial r} + \frac{U_r}{r} + \frac{1}{r} \frac{\partial U_\varphi}{\partial \varphi} \; = 0 \, .$$

At first we investigate the solution of the system of equations (4.6)–(4.8), (4.40) neglecting the dependence of E on r in the formula (4.2); the effect of this dependence will be studied below. The solution of the problem in the cilindrical coordinates is carried out similarly to Sec. 4.1.2 with the substitution in the expressions (4.10) $\exp^{iqx} \longrightarrow J_m(qr) \cos m\, \varphi$ where J_m is Bessel function of the first order, m is the integer. As a result, one obtains for the components of the displacement vector

$$U_z = (qN e^{-\kappa_t z} + M\kappa_1 e^{-\kappa_1 z} - \lambda^2 M \sum_{j=n,T} \Phi_j e^{-\delta_j z}) \, J_m(qr) \cos m\, \varphi \, e^{\lambda t} \, ,$$

$$U_r = (-\kappa_t N e^{-\kappa_t z} - qM e^{-\kappa_1 z} + \lambda^2 M \sum_{j=n,T} (q/\delta_j) \Phi_j e^{-\delta_j z}) \; .$$

$$\frac{1}{2} (J_{m-1}(qr) - J_{m+1}(qr)) \cos m\, \varphi \, e^{\lambda t} \, ,$$

$$U_\varphi = (\kappa_t N e^{-\kappa_t z} + qM e^{-\kappa_1 z} - \lambda^2 M \sum_{j=n,T} (q/\delta_j) \Phi_j e^{-\delta_j z}) \; .$$

$$\frac{1}{2} (J_{m-1}(qr) + J_{m+1}(qr)) \sin m\, \varphi \, e^{\lambda t} \, . \tag{4.41}$$

From (4.40) and (4.41) one obtains the dispersion equation of EDTI, which exactly coincides with the equation (4.22). So the formulae for q_m and λ_m (4.28), selecting the dominant structure in the linear regime, are valid also in the case of the structures with the axial symmetry.

As it is seen from (4.22) the surface harmonics with an arbitrary m have the same growth rate. This "m-degeneracy" is the consequence of the neglect of the dependence of the intensity of the laser field $|E|^2$ on r in the formula (4.2).

If the intensity of the laser field has the Gaussian distribution

$$E^2 = E_0^2 \exp(-r^2/r_0^2), \tag{4.42}$$

then this "m-degeneracy" is removed (see [307]). For example in the case of radial-ray structure with m rays, when

172

$$\xi = A \left(\frac{r}{r_0}\right)^m e^{-r^2/r_0^2} \cos m\varphi\, e^{\lambda t}$$

$$Y_{j1} = A_j \left(\frac{r}{r_0}\right)^m e^{-r^2/r_0^2} \cos m\varphi\, e^{\lambda t}, \tag{4.43}$$

the growth rate $\lambda = \lambda(m)$ attains maximum value for

$$m_{max} = \frac{q_{R\epsilon}^2\, r_0^2}{4}\, [(5a^2 + 6a + 1)^{1/2} - (2a + 1)]^{1/2} \tag{4.44}$$

The value m_{max} — (4.44) determines the number of the rays in the dominant structure.

4.1.5. The comparison with the experiment and conclusions

As it already has been pointed out, the symmetry of the structures, formed as a result of EDTI development, must be determined by the symmetry of the laser beam in the case of the isotropic surface, or by the symmetry of the surface of the crystal. Let us estimate the parameters and make the comparison of the theoretical conclusions with the experimental results. For the characteristic values $\theta = 10^{-11}$ erg, $\rho = 5$ g/cm^3, $c_1 = 5 \cdot 10^5$ cm/s, $\beta = 0{,}4$, $\omega = 4 \cdot 10^{15}$ c^{-1}, $c_v = 2 \cdot 10^7$ erg/cm^3, $K = 10^{12}$ erg/cm^3, $a = 2 \cdot 10^{-5}$ K^{-1}, $\partial\epsilon''/\partial\omega = 10^{-15}$s, $|\partial E_g/\partial T_0| \cong 4 \cdot 10^4$ eV/K, $|\partial E_g/\partial n_0| = 10^{-32}$ erg/cm^3, $\chi = 0{,}1$ cm^2/s, $D = 10^2$ cm^2/c, $\gamma_0 = 10^5$ cm^{-1}, $\gamma_A = 4 \cdot 10^{-31}$ cm^6/s, we have from (4.17) $R_T \sim R_n \sim \epsilon_{TT} \sim \epsilon_{nn} \sim (10^4 - 10^5)\, |E|^2$. From (4.33) one obtains for the critial intensity of DTI $I_{Tc} = 2 \cdot 10^6$ W/cm^2, for $q = 10^4$ cm^{-1}. The critical intensity of EDI $I_{nc} \sim D/\chi\, \hbar\omega |\partial E_g/\partial T_0| / c_v |\partial E_g/\partial n_0| \sim 10\, I_{Tc}$. From (4.28) in the case of EDI we obtain $q_m \sim q_{RE} \sim 10^{-2} |E|^2 \cong (10^4 - 10^5)$ cm^{-1}, for $I = 6 \cdot 10^8$ W/cm^2 the period of the structures $d = (1 \div 5)\, \mu$m. From (4.29) we obtain for the growth rate $\lambda_m \sim Dq_m^2 \sim 10^{10} \div 10^{11}$ s^{-1}. The amplitude of the renormalization of the forbidden gap in (4.38) $E_{g1} \sim 1$ eV, i.e. EDTI can lead to the semiconductor-metal phase transition. This conclusion and the estimates of the periods of the structures and their growth rates are in accordance with the experimental results [280]. The formation of the semiconductor-metal periodic structures (gratings of "the sun"-structures, depending on the symmetry of the laser field) was observed on the isotropic surface of VO_2. According to the formula (4.44) we have in the case $r_0 = 5 \cdot 10^{-3}$ cm [280] for the number of the rays in the "sun" structure $m \sim 50$, in accordance with the experiment.

In the experiment [284] the formation of one dimensional gratings of the surface relief was observed after the irradiation of the (111) surface of the crystalline Si by the laser pulses with the wavelength $\lambda = 0,53$ μm. The fronts of the gratings were directed along 110 axis. The dependence of the grating orientation on the crystallagraphic directions points out on the deformational character of the mechanism of their formation. The period of the gratings did not depend on λ, the angle of incidence or polarization of the incident radiation, and was determined by the duration of the pulse. For $\tau_{1p} = 10^{-11}$ sec, the period $d_1 = 3 \cdot 10^{-5}$ cm, and for $\tau_{2p} = 5 \cdot 10^{-12}$ s, $d_2 = 6 \cdot 10^{-6}$ cm, that is $d_1/d_2 \sim (\tau_{1p}/\tau_{2p})^{1/2}$. For Si we have $\epsilon_{nn}/R_n \sim 10 \gg 1$, so the period of the gratings must be determined by the formula (4.30) that is $d \sim \sqrt{\tau_p}$ in accordance with the experiment [284].

We note, that in the works [70, 279], the mechanism of the formation of the supergrating with the alternation of the metallic and dielectric phases owing to the laser-induced phase transtion was considered. The mechanism of the EDTI and of the corresponding laser-induced phase transition, as well as the class of superstructures which are formed as a result of EDTI development, are qualitativly different from that considered in [70, 279]. The supergrating [70, 279] is generated as a result of the interference of the incident light wave with the wave, difracted on the initial grating of the dielectric permeability, which is formed owing to the fluctuating Fourie-harmonies n_1 and T_1 in (4.1) (the deformations are not included into the consideration). On the contrary under EDTI the long range order is estiblished owing to the surface deformation of the medium and the characteristic periods of the structures are not determined directly by the wavelength of the laser radiation.

As it was shown here, besides the statical structures, under the EDTI the stimulated generation of SAW (dynamical surface structures) occurs. We note that in the work [98] the threshold generation of the nonthermal SAW under the action of the laser radiation on the GaAs surface was recorded. The estimates of the threshold intensity according to the formula (4.24) yields $I_c \sim 10^4$ W/cm^2 in accordance with the experimental value [98]. More detailed study of the generation of SAW under EDTI was carried out in the work [99].

The mechanism of EDTI, considered in this section can be interesting also for the problem of ultrafast laser melting of the surface [126, 292] (which under EDTI can occur owing to the SAW frequencies softening, generation of the point defects [285] and other problems of the pulsed laser-solid interactions.

174

4.2. Laser-induced defect-deformational instabilities and formation of the ordered defect structures

The role of the diffusional variable under DDI can be played by the concentration of the different defects in solids. In this section we briefly consider two examples of the endothermal DDI under the action of the laser field with the formation of ordered structures of the dislocations and the voids. The analogous DDI can develop with the participation of vacancies [290].

4.2.1. Dislocational-deformational instability and formation of the dislocational gratings under the action of laser light

Under the laser action on the surface of the crystal, owing to the spatial nonuniformity of the absorbed energy across laser beam and into the depth of the medium, the shear thermoelastic deformations arise, which lead to the generation of the dislocations [286]. Usually one considers the laser-induced dislocations to be randomly distributed in the crystal. In this section we shall show, that when the shear stress exceeds some critical value $\tau = \tau_c$ the dislocational-deformational instability (DIDI) develops under which the dislocations glide and form spatially periodical field of the dislocation concentration coupled with the deformations. The Fourier-amplitude of this field plays the role of the order parameter of phase-transition of the second order.

We note that the formation of the gratings of the dislocation density on the Si surface under the action of the millisecond Nd-laser pulses was observed in Ref. [287]. The estimates of the parameters of DID-structures show, that DIDI can be responsible for this effect. The formation of the periodical dislocational gratings is observed also under the repeated pulsed mechanical action on the surface of the metals [288]. This effect play the important role in the process of the fatigue fracture of the material. Analogously one can expect, that the formation of DID-structures under laser action, can play the important role in the process of the surface damage. The DIDI effect can be interesting also for the problem of the laser mirror reflectivity, supergrating formation and other problems of laser modification of the surface.

Let the plane (x, y) be a gliding plane of the edge dislocations (consider for example the surface (111)). This surface is acted upon by the laser beam with radius r_0, which leads to the generation of the edge dislocations in the subsurface layer. The dislocations move in the plane (x, y) along the three directions [100]. Let one of these directions coincides with the axis x. The edge dislocations are extended along the y-axis and their Burgess vectors are directed along x-axis (Fig. 45).

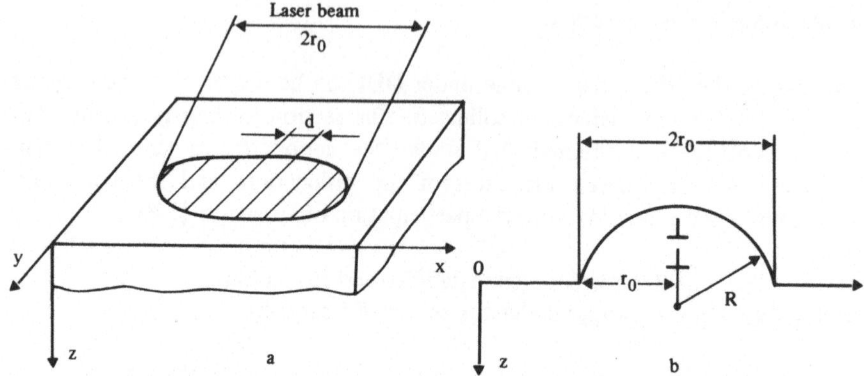

Fig. 45. The geometry of the dislocation grating formation. a) **b** – Burgess vector of
the edge dislocation, d – the period of dislocation gratin; b) The relief of
the deformed surface.

Let us introduce the density of the number of the dislocations with Burges
vectors **b** ‖ x: $\rho(x, z) = \sum_i \delta(\lambda - x_i)\delta(z - z_i)$, where the summation is carried out
over all dislocations with **b** ‖ x. The movement of these dislocation along x-axis
owing to the diffusion [288] and the drift is described by the equation:

$$\frac{\partial \rho}{\partial t} = D \frac{\partial}{\partial x^2} \rho - \frac{\partial}{\partial x} v \rho \tag{4.45}$$

Here D is the diffusion coefficient of the dislocations, v_m is the drift velocity
[289]:

$$v = v_x = M\epsilon^m \tag{4.46}$$

where M and m are the constants (> 0), ϵ is the shear deformation

$$\epsilon = \partial U_z / \partial x . \tag{4.47}$$

The equations (4.45)–(4.47) are closed by the equation for the vector of the displacement of the medium. In the approximation of the isotropic medium,
in the geometry considered ($U_x = 0$, $\partial^2 U_z / \partial z^2 = 0$) it reads

$$\frac{\partial^2 U_{1z}}{\partial t^2} = c_t^2 \frac{\partial^2 U_{1z}}{\partial x^2} + c_t^2 \beta_a \frac{\partial^2 U_{1z}^3}{\partial x^2} + bc_t^2 \rho , \tag{4.48}$$

176

where $b = |b|$, ρ is the constant of anharmonicity of the elastic continuum. Now we perform the linear stability analysis of the system (4.45)–(4.48).

The threshold of DIDI and the growth rates. We represent the variables of the problem in the form $\rho = \rho_0 + \rho_1$, $v = v_0 + v_1$, $\epsilon = \epsilon_0 + \epsilon_1$ where ρ_0, ϵ_0, v_0 are the mean, slowly varying values, and ρ_1, v_1, ϵ_1 are the fluctuating diviations. From (4.45)–(4.48) one obtains the equations for the fluctuations, which for the geometry considered here, have the form

$$\frac{\partial^2 U_{1z}}{\partial t^2} = c_t^2 \frac{\partial^2 U_{1z}}{\partial x^2} + b\rho_1 , \tag{4.49}$$

$$\frac{\partial \rho_1}{\partial t} = D \frac{\partial^2}{\partial x^2} \rho_1 - g \frac{\partial^2 U_{1z}}{\partial t^2} , \tag{4.50}$$

where the constant of the coupling between the deformation field and the dislocation field:

$$g = mM\rho_0 \epsilon_0^{m-1} = m\rho_0 \frac{v_0}{\epsilon_0} \tag{4.51}$$

is determined by the external pump. We seek the solution of the system of equations (4.49), (4.50) in the form

$$U_{1z} = A_q \exp(iqx + \lambda t)$$

$$\rho_1 = B_q \exp(iqx + \lambda t) , \tag{4.52}$$

where A_q, B_q are the initial amplitudes. Substituting (4.52) into (4.49) and (4.50), one obtains two equations for A_q and B_q, whence the expession for the growth rate follows $(q^2 c^2 > \lambda^2)$

$$\lambda = gb - Dq^2 = m\rho_0 \frac{v_0}{\epsilon_0} b - Dq^2 . \tag{4.53}$$

For $gb \geqslant Dq^2$ the growth rate $\lambda \geqslant 0$ and DIDI starts to develop. Thus, the threshold value of the externally varied parameter depends on q and is equal to

$$(gb)_{th} = mMb(\rho_0 \epsilon_0^{m-1})_c = Dq^2 \tag{4.54}$$

The nonlinear stationary regime and the formation of the dislocational-deformational grating.

12 Akhmanov, Laser

177

For the determination of the period of DID-structures, formed as a result of DIDI, we, from (4.48), (4.50), using (4.52), where $\lambda = 0$, find the equation for the stationary amplitude

$$-q^2 A_q - \beta q^2 |A_q| A_q + \frac{bg}{D} A_q = 0 , \qquad (4.55)$$

whence

$$|A|_q = \beta^{-1/2} (\frac{bg}{Dq^2} - 1)^{1/2} , \qquad (4.56)$$

where $q \leqslant q_0 = (bg/D)^{1/2}$.

The resulting DID-structure is determined by the summation over the Fourie-harmonics with the vector q. Taking into account (4.52), where $\lambda = 0$ and (4.56) we obtain for the component of the displacement vector

$$U_z = \sum_{|q| \leqslant q_0} A_q \exp iqx \sim \int_0^{q_0} q (\frac{bg}{Dq^2} - 1) \cos qx \, dq \sim J_1(q_0 x) \quad (4.57)$$

where J_1 — is Bessel function. The deformation $\partial U_z / \partial x \sim J_0(q_0 x)$.

The period of the stationary DID-grating is determined by the expression

$$d \cong \frac{2\pi}{q_0} = 2\pi (\frac{D}{bg})^{1/2} \qquad (4.58)$$

To find the value of the period of the grating with the help of the formula (4.58) we estimate the values of ρ_0 and ϵ_0 for the conditions of the experiment [287], where the radius of the laser beam $r_0 \cong 10^{-2}$ cm. According to [287], the deformed surface relief of Si inside the laser spot can be approximated by the semisphere with the radius $R \cong 2$ cm (see Fig. 46). Then $\epsilon_0 = \partial U_{z0} / \partial x \sim$ $\sim U_{z0}/r_0 \cong (R - \sqrt{R^2 - r_0^2})/r_0 \approx r_0 /2R \cong 10^{-3}$. To estimate ρ_0 we use [286]: $\rho_0 = (Rb)^{-1} \sim 10^8$ cm^{-2} for $b = 10^{-8}$ cm. For the velocity of the movement of the dislocation we have $v_0 \cong 10^{-3}$ cm^{-1} [289] (Si, the tempera-ture $T \cong 10^3$K, the shear stress $\sigma_0 = \mu\epsilon_0 \cong 10^9$ din/cm^2 for $\mu = 10^{12}$ din/cm^2). Then for $m > 1$ from (4.51) we have $g \approx 10^8$ cm^{-1}c^{-1}. Let us now estimate the diffusion coefficient for the edge dislocation considering the latter as a cilinder with the radius b and the length $l \cong 10^{-4}$ cm, the density $\rho \cong 2$ g/cm^3, and the mass $m \cong 5 \cdot 10^{-19}$ g. The diffusion of the dislocation can be considered as a rondom movement of this cilinder in the one dimensional Paierls relief with

178

the activation energy W. The diffusion coefficient of one atom with the mass m_a in this relief is $D_a = a^2\nu_a \exp(-W/kT)$, where $\nu_a = \text{const } m_a^{-1/2}$, a is the lattice parameter. The frequency of the "jumps" of the dislocation $\nu_d = \text{const}/m_d^{1/2}$, thus $D_d = (m_a/m_d)^{1/2}D_a \cong 10^{-2}D_a \sim 10^{-7}\text{cm}^2\text{c}^{-1}$ for $D_a \cong 10^{-5}$ cm^2/c (for premelting temperature). Then from (4.58) one has $d \cong 10^{-4}$ cm in accordance with the experimental value of Ref. [287].

In our consideration we used the approximation of the isotropic medium. The anisotropy of the crystal in this simplest model is taken into account by the selection of the axis of the easy gliding (x-axis), along which the wave vector **q** is directed. In the plane (111) there are three equivalent axis of the easy gliding directed at angles 60° to each other. Thus under the irradiation of (111)-surface three gratings with the wave vectors **q** forming the equilateral triangle must be generated. On the surface (100) there are two such axis, thus here two mutually ortogonal gratings must form, which in fact was observed in the experiment [287].

4.2.2. The laser induced void-deformational instability and formation of the ring structures of the voids in the thin films.

In the thin metallic films deposited on the dielectric substrates the vacancies form the voids [291]. If the thin film is acted upon by the laser beam with Gaussian intensity distribution across the beam then the void-deformational instability (VODI) can develop, which leads to the formation of the ringed void structures, firstly observed in [293]. VODI is described by the linearized equation for the coordinate of bending deformations of the film [158], interacting with the voids

$$\frac{\partial^2 \varsigma_1}{\partial t^2} + \frac{h^2 c^2}{12} \Delta_r^2 \varsigma_1 - c_{11}^2(r)\Delta_r \varsigma_1 = -RN_1; \quad R \equiv \sigma_\perp S_0/h\rho , \qquad (4.59)$$

where h is the thickness of the films, Δ_r is the lateral (along the film) Laplace operator, c is the constant of the order of the sound velocity, $c_{11}^2 = \sigma_{11}(r)/\rho$, σ_{11} is the lateral stress and $\sigma_\perp$ is the normal to the film stress; both arise due to the difference of the thermal expansion coefficients of the film and substrate, S_0 is the area of the foundation of cylindrical void with the height h, $N_1(\text{cm}^{-2})$ — the fluctuation of the number density of the voids ($N = N_0 + N_1$, N_0 — the mean value of number density of voids). The equation for N_1 has the form [293]:

$$\frac{\partial N_1}{\partial t} = D\Delta_r N_1 + g\Delta_r^2 \varsigma_1 , \qquad (4.60)$$

where $D = D_{v0}a_0^3 n_{v0}$ – the diffusion coefficient of the voids, D_{v0} is the coefficient of lateral diffusion of vacancies in the film, a is the lattice parameter in the film, n_{v0} – the concentration of vacancies, $g = -hDN_0 |\theta_v|/4kT_0$, $|\theta_v| = Ka^3$, K is bulk elastic modulus. Taking the solution of (4.59), (4.60) in the form $N_1, \zeta_1 \sim J_0(qr)\exp \lambda t$, we obtain for the growth rate

$$\lambda = -Dq^2 + \frac{bq^2}{1 + l_0^2 q^2} \, , \tag{4.61}$$

where $b = Rg/c_{11}^2$, $l_0^2 = (c^2/c_{11}^2)h^2/12$. The maximum value $\lambda = \lambda_m$ is atained in the point $q = q_m$:

$$\lambda_m = \frac{2}{l_0^2} \left(\frac{b + D}{2} - \sqrt{bD} \right) , \qquad q_m = \frac{1}{l_0} \left(\sqrt{\frac{b}{D}} - 1 \right) \tag{4.62}$$

and the period of the concentric ringed structures of the void concentration and bending deformation field is given by the expression

$$d \cong \frac{2\pi}{q_m} = \frac{2\pi l_0}{\sqrt{b/D - 1}} \, . \tag{4.63}$$

The formation of such ringed structures of the voids is, in fact, observed under laser deposition of the metal films on the dielectric substances. More detailed exposition of the VODI theory and experimental results are given Ref. [293].

4.3. Crystallizational-deformational-thermal instability and the formation of the ordered crystalline-amorphous structures under laser crystallization

Besides the endothermical DDI considered above, the exothermical DDI are possible, which develop due to the energy hidden in the medium, for example, the latent crystallization heat in amorphous medium. The pulsed laser action in this case plays the role of initial activation. We shall consider this type of DDI on the example of the crystallizational-deformational thermal instability (CDTI), occuring on the amorphous surface of semiconductors [305].

As has been already pointed out in Introduction and Sec. 1.3 the processes of the crystallization of the amorphous semiconductors under the action of the pulsed laser radiation (with durations $\tau_p \sim 10^{-8} - 10^{-11}$ s) are widely investigated at present. The interesting feature of pulsed laser crystallization is the formation of the periodical concentric ring structures, with the multiple alternation of crystalline and amorphous regions (the number of the rings can be as big as five [287] or eight [294]). Under the irradiation of the amorphous germanium films the appearance of more complicated periodical structures was observed which are formed by the concentric rings in the center of the laser spot and the radial rays in the outside region ("the sun" [305], or two-dimensional grating [274]). The occurence of these periodical structures points out to the development of some instability on the irradiated surface.

In this section a new type of laser-induced instability is considered − the cristallizational-deformational-thermal instability (CDTI), which occurs on the amorphous surface of the semiconductors. The theory of CDTI can explain the formation of the mentioned above structures. The physical mechanism of CDTI consists in the following. The amorphous phase is a metastable energetically exited state with the latent heat of the crystallization E_k. The rate of crystallization at the room temperatures and in the absence of the deformation is negligibly small. Laser heating up to the temperature T_0 gives rise to the local increase of this rate (see (4.64)). The fluctuating Fourie-harmonic of the surface deformation modulates the crystallization rate, in accordance with the formulae (4.64), and (4.67). This leads to the modulation of the latent heat liberation and, owing to this, to the spatial modulation of the temperature (see (4.68)). Owing to the spatially nonuniform (along the surface) temperature field, the thermoelastic force arises ($F \sim \mathrm{grad}\, T_1$), which amplifies the initial deformation. This positive feedback under the condition $T_0 > T_c$ (where T_c is the critical temperature) leads to the instability, in which the Fourie-amplitudes of the temperature perturbations, statical deformation as well as the crystallization rate grow exponentially in time. The CDTI develops under the condition that the meltings dose not occur. We must stress the fact, that the impulsive laser heating plays only the role of the initial activation, and the instability develops owing to the latent heat of the crystallization. This constitutes the principle difference between the CDTI and EDTI (Sec. 4.1): the latter leads to the formation of the analogous deformational structures, but develops owing to the absorbed energy of the laser radiation. On the other hand, the decisive role of the deformation distinguish CDTI from a tratitional explosive crystallization, which develops owing to the thermal effects only [295].

As a result of the CDTI the concentric ring fields of the temperature and deformation are formed on the surface as well as the resulting ring field of alternations of the amorphous and crystalline phases ("the sun" − structures as well as the gratings can also be formed). The estimates of the parameters of these structures shows that the theory of CDTI can explain the experimentally observed periodic structures.

We must note, that besides the sophisticated ordered structures, more simple formations are observed under laser irradiation (in the main by the picosecond pulses) of the amorphous semiconductors, consisting in crystallihe spot in the center and one crystalline ring with the dimension of the order of that of the laser spot [276, 277]. The dynamics of the development of these structures shows that their formation takes place not as a result of an instability, but owing to the effects arising due to the Gaussian distribution of the intensity across the laser beam.

Under the certain conditions the CDTI, considered in this section, may arise not only under laser action but also under other methods of heating, for example, by the electron beam, or even in the furnace. It is interesting to note, that in the latter case the cencentric rings and more sophisticated periodical structures of the alternation of the different phases are observed [296].

The equations, describing the crystallization, heating and deformation. Let us consider the amorphous medium, occupying the halvespace $z \geqslant 0$. Let $a = a(x, y)$ $(0 \leqslant a \leqslant 1)$ is the local relative volume, occupied by the amorphous phase $a = \Delta V_a / \Delta V$, $\Delta V_a + \Delta V_c = \Delta V$, where ΔV is the physically small volume, ΔV_a, ΔV_c are correspondingly the volumes of the amorphous and crystalline phases. The quantity a is determined by the activational kinetic equation

$$\frac{\partial a}{\partial t} = -\frac{a}{\tau_0} \exp\left(-\frac{W + \theta_a \operatorname{div} U}{k_B T}\right).$$

(4.64)

Here W is the activation energy, τ_0 — is the phenomenological constant, θ_a is the deformational potential. The equation (4.64) with $\theta_a = 0$ is usually used for the description of the thermal explosive crystallization [295]. For $\theta_a \neq 0$, the equation (4.64) takes into account the fact, that the deformation influences the rate of the crystallization.

The equation for T has the form

$$\frac{\partial T}{\partial t} = \chi \Delta T + \frac{E_k}{c_v \tau_0} a \exp\left[-\frac{W + \theta_a \operatorname{div} U}{kT}\right] + \frac{c\gamma |E|^2}{2\pi c_v} e^{-\gamma z},$$

(4.65)

where E_k is the latent heat of the crystallization. In equation (4.65) we have neglected the change of the value of E_k under the deformation: $E_k = \Delta\theta \cdot \operatorname{div} u$, where $\Delta\theta$ is the difference of the deformation potentials in the crystalline and the amorphous phases.

The boundary conditions are given by equation (4.6). The equations (4.64) and (4.65) are closed by the equation for the displacement vector U — (4.7), where now $f_n = 0$, $Y_{T1} = T$ (we neglect the difference of the densities of the crystalline and amorphous phases). The boundary conditions are given by the equations (4.40).

Take $T = T_0 + T_1$, $a = a_0 + a_1$, $\xi = \operatorname{div} U = \xi_0 + \xi_1$, where T_0, a_0, ξ_0 are the spatially uniform (along the surface) solutions and T_1, a_1, ξ_1 are small spatially nonuniform perturbations. Linearising the equations (1) and (2), we obtain the equations of zeroth order

$$\frac{da_0}{dt} = -\frac{a_0}{\tau}\ ; \qquad \tau^{-1}(z) = \frac{1}{\tau_0}\exp\left(-\frac{W_0}{k_0 T_0}\right), \qquad W_0 = W + \theta_a \xi_0, \tag{4.66a}$$

$$\frac{dT_0}{dt} = \chi \Delta T_0 + \frac{E_k}{\tau c_v}a_0 + \frac{c\gamma E^2}{2\pi c_v}e^{-\gamma z}. \tag{4.66b}$$

and the equations of the first order

$$\frac{da_1}{dt} = -\frac{1}{\tau_0}e^{-W_0/kT_0}\left[a_0\left(-\frac{\theta_a}{kT_0}\xi_1 + \frac{W_0}{kT_0^2}T_1\right) + a_1\right], \tag{4.67}$$

$$\frac{dT_1}{dt} = \chi \Delta T_1 + \frac{E_k}{\tau_0 c_v}e^{-W_0/kT_0}\left[a_0\left(\frac{W_0}{kT_0^2}T_1 - \frac{\theta_a}{kT_0}\xi_1\right) + a_1\right]. \tag{4.68}$$

Note, that if one neglects the thermal conductivity ($\chi = 0$) then from (4.67) and (4.68) one obtains the law of the energy conservation $d/dt(E_k^q + C_v T_1) = 0$.

For the full description of CDTI it is necessary to solve the nonlinear equations (4.66b) (with taking into account of (4.66a)), which is the separate problem. For our purposes it will be sufficient to make certain assumptions about the form of this solution. We expand $\xi_0(t, z)$ in power series $\xi_0^{(z)} = \xi_0^{(0)} - \xi_0' z$, $\xi_0' \equiv |\partial \xi_0/\partial z|$ and use this expansion in (4.67) and (4.68). Then we obtain

$$e^{-\dfrac{W + \theta_a \xi_0}{kT_0(z)}} \cong e^{-\dfrac{W_0(0)}{kT_0(0)}}\ e^{-\dfrac{|\theta_a|\xi_0'}{kT_0(0)}z} \tag{4.69}$$

The approximation (4.69) is valid under the condition $\theta_a/kT_0(0) \gg 1$. In addition, under the condition $\lambda\tau \gg 1$, one can neglects the term with a_1 in the right hand side of (4.67), (4.68). Then the equation (4.68) with taking into account of (4.69), can be written in the form

$$\frac{\partial T_1}{\partial t} = \chi \Delta T_1 + \sigma e^{-\Gamma z}T_1 + g\operatorname{div} U\, e^{-\Gamma z}, \tag{4.70}$$

183

where

$$\sigma = \frac{1}{\tau(0)} \left[\frac{E_k}{c_v T_0(0)} \right] \left[\frac{W_0(0)}{k T_0(0)} \right] a_0 \equiv S \frac{e^{-W_0(0)/k T_0(0)}}{k^2 T_0^2(0)} , \qquad (4.71a)$$

$$g = -\frac{T_0(0)}{\tau(0)} \left[\frac{E_k}{c_v T_0(0)} \right] \left[\frac{\theta_a}{k_{\text{Б}} T_0(0)} \right] a_0 ,$$

$$\Gamma = \frac{|\theta_a|}{k_B T_0(0)} |\partial \xi_0 / \partial z| . \qquad (4.71b)$$

The equations (4.70) and (4.6), now represent the closed system of equations for T_1 and U, the solution of which determines the first order perturbation of the crystallization rate, in accordance with (4.67).

The formation of ordered structures of different phases. If the change of T_1 and divU with the distance from the surface is small on the length of order Γ^{-1}, then one may put in the last two terms of the right hand side of (4.70) $T_1(z) = T_1(0)$, $(\text{div}U)_z = (\text{div}U)_{z=0}$ and write down this equation in the form

$$\frac{\partial T_1}{\partial t} = \chi \Delta T_1 + [g (\text{div}U)_{z=0} + \sigma T_1(0)] e^{-\Gamma z} . \qquad (4.72)$$

The equations (4.72) and (4.4) form the closed system of equation, which is' analogous to that of EDTI (Sec. 4.1). The analysis of the equations (4.17) and (4.4), performed in Sec. 4.1 is valid also in the case of CDTI. Thus, in analogy with Sec. 4.1, the following possible types of surface structures can be formed as a result of CDTI development: 1) concentric rings, "sun" structures and cells, formed in the point of intersections of the rings and the radial rays. The surface deformation and temperature fields in this case have the form (compare (4.41)): 1) T_1, $(\text{div}U)_{z=0} \sim J_m(qr) \cos m\varphi \, e^{\lambda t}$; 2) radial rays (compare (4.44)) T_1, $(\text{div}U)_{z=0} \sim r^m \cos m\varphi \, e^{\lambda t - r^2/r_0^2}$; 3) one-dimensional grating (4.10) T_1, $(\text{div}U)_{z=0} \sim e^{iqx + \lambda t}$. These surface fields, as can be seen from (4.67) modulate the crystallization rate. Thus, as a result of CDTI the above considered types of periodic structures are formed on the surface by the alternation of the amorphous and crystalline phases. The type of the structure depends on the symmetry of the surface, or that of the laser field.

184

The dispersion equation for CDTI, determining the dependence of the growth rate λ on wave number q for various values of the parameters of the problem, is quite analogous to the dispersion equation of EDTI (4.22), and can be written in the form (it directly follows from (4.22), where j = T):

$$(\kappa_t^2 + q^2) - 4\kappa_t\kappa_1 q^2 = R\,\frac{\lambda^2}{c_l^2\,\chi\Gamma\delta}\,\frac{(\kappa_t^2 + q^2)^2 - 4\kappa_t\delta q^2}{(q^2 - \delta^2)(1 - (R + \sigma)/\chi\Gamma\delta)}\,, \tag{4.73}$$

where the quantity

$$R = \frac{Ka E_k\,\theta_a\,a_0}{\rho c_l^2\,\tau_0\,c_v\,k_B T_0(0)}\,\exp\left(-\frac{W}{kT_0(0)}\right) = \frac{Qe^{-\frac{W}{kT_0(0)}}}{kT_0(0)} \tag{4.73}$$

reflects the coupling (through deformational potential θ_a) of the crystallizational $\left(\dfrac{E_k}{\tau_0}e^{-W/k_B T_0(0)}\right)$ and deformational (Ka) characteristics of the system and is controlled by the externally variable parameter $T_0(0)$.

The solutions of the equation (4.73) is given by the formula (4.33), (4.27) where one must replace $\chi_j \to \chi$, $\tau_{j0}^{-1} = 0$, $\gamma_0 \to \Gamma$, $R_j \to R$, $\epsilon_{jj} \to \sigma$, $q_R \to R/\chi\Gamma$, $q_{R\epsilon} \to q_{R\sigma} = (R + \sigma)/\chi\Gamma$. In the case $R \gg \sigma$ maximum value of the growth rate $\lambda = \lambda_m$ and the value of q for which this value is reached (q $= q_m$) are given by the formula (4.28). Thus the period of the structure, formed owing to the CDTI, is given by the formula $d \cong 2\pi/q_m$. For the $\sigma \gg R$ (i.e. $q_{R\sigma} \gg q_R$) the maximal value λ_m is reached in the point q = 0. Considering the nonlinear regime of CDTI (adding the anharmonic term to the equation for the displacement vector **U**) one can show, that in this case the structure with $q_m \cong q_R$ is formed with the period $d \cong 2\pi/q_R$ (compare (4.31)).

The critical condition on the externally variable parameter $T_0(0)$ can be obtained from the condition $\lambda = 0$. Then from (4.33) making the above mentioned substitution and using (4.73) one has

$$\frac{e^{-W_c}}{W_c}\,Q + \frac{S(1 - \beta)}{W_c} = \chi\Gamma q(1 - \beta)\,;\qquad W_c = \frac{W}{k_B T_c} \tag{4.74}$$

Thus, from the above consideration it follows, that under the impulsive laser heating $(T_0(0) > T_c)$ CDTI develops on the initially amorphous surface with the appearance of ordered surface structures formed by the coupled fields of the temperature, deformation of the medium and periodic alternations of different phases.

Let us estimate the period of the structures in the case of laser pulse action on the amorphous Si-films. The characteristic values of the a-Si are $\tau_0 = 3.2 \cdot 10^{-15}$ sec, $E_K = 10^3$ J/cm^3 [295], $\theta_a = 10$ ev, $C_v = 2$ J/cm^{3}K, $\rho_2 = 2.3$ g/cm^3, $\beta = 0.2$, $K = 10^{12}$ erg/cm^3, $a = 10^{-6}$K^{-1}, $\chi = 0.1$ cm^2/sec; take for the estimate $T_0(0) = 1400$K (the melting temperature of a-Si). Then in (4.73) we have $Q = 3.3 \cdot 10^{14}$ sec^{-1}, $R = 4 \cdot 10^6$ sec^{-1}, for $W = 1.9$ eV. From (4.71a) we have $S = 4.2 \cdot 10^{17}$ sec^{-1}, $\sigma = 3.3 \cdot 10^8$ sec^{-1} (the value of the activation energy depends on the character of the amorphisation and varies in the limits $W = 1.8 \div 2.3$ eV [297–299]. The value of Γ – (4.76b) depends on values of thermal conductivity, optical absorption, the pulse duration τ_p. For $\Gamma = 10^5$ cm^{-1}, $\sigma \gg R$ we obtain $q_m = 2 \cdot 10^4$ cm^{-1}, $d = 3$ μm, $\lambda_m = 7 \cdot 10^7$ sec^{-1}, $/\tau^{-1}(0) = 6 \cdot 10^7$ sec$^{-1}/$.

From (4.74) we have $W_c = 16$ for $W = 1.9$ eV, $q = 2 \cdot 10^4$ cm^{-1}, $\Gamma = 10^5$ cm^{-1}, and $W/K_B T = 15.5 < W_c$ for $T = 1400$K. Thus, the critical condition (4.74) can be fullfield in the conditions, considered above.

The symmetry of the observed structures depends on the symmetry of boundary conditions on the surface. For examples in the work [306] the axial "sun structure" was observed, while in Ref. [274] the observation of the two-dimensional grating was reported. The periods of these structures (~ 1 μm) corresponds to the results of CDTI-theory duscussed above.

CONCLUSIONS.

1. Let us summarise our discussion. The modern laser technique having advanced into the femtosecond range of generated optical pulse durations, practically solved the problem of exitation and recording of nonquilibrium states in condensed media and has reached the natural limit of the rate of optical response of such media, determined by one optical field cycle duration.
In the case of semiconductors and metals this technique enables injecting extremely rapidly (over a time 10^{-14}s) significant energy into the electronic subsystem of the surface layers and afterwards following in details the various stages of energy relaxation to the equilibrium lattice state and the subsequent cascade of phase transformations on the surface. The scale of relaxation processes in solids acsessible to the experimental investigation covers wide range from the ultrafast electron-electron dephasing and relaxation energy processes (with characteristic duration of 10^{-14}s) electron-phonon relaxation through LO-phonons ($10^{-14} - 10^{-13}$s) and TA-phonons ($10^{-13} - 10^{-9}$s) emission right up to the interband recombination processes in semiconductors and various phase transformations and the processes of heat and mass conduction with the characteristic times in nanosecond range and longer.
2. The PLA experiments in nanosecond and picosecond range which arosed much interest and debates are now received the qualative expanation in the terms of described above scale of relaxation processes with the use of "thermal" model. According to it under the action of sufficient short and powerful laser pulse with the energy quantum exceeding the forbidden energy gap, the light energy absorbed by the electron-hole subsystem of crystal in less then 1 ps is transfered from the exited above the conduction band bottom free curriers to the crystal lattice and melts it. The melting front moves with the velocity of the order of velocity of sound into the medium and after the end of the laser pulse stops at certain depth and afterwards in nanoseconds it returns back to the surface, followed by the epitaxial regrowth of crystal lattice. From several problems referring to PLA remained unsolved one must once more points out the question about the degree of equilibriuty (or nonequilibriuty) of phonon subsystem at the beginning of the melting[1] and also the possibility of new types of phase transitions caused by the superdence plasma of free carriers on the first stage of PLA (on the time intervals < 1 ps) [60]. Another

---

[1] We note that recently the possibility of direct registration and spectroscopy of shortwave acoustical phonons generated by the action of short (including pico- and subpicosecond) laser pulses on the semiconductor surface [195–198].

model of "cold" melting was proposed in Ref. [190] (see also Sec. 4.1.). The development of EDTI leads to the large statical displacement of the atoms from their equilibrium positions, so that the laser pulse energy is transformed into the potential energy of the lattice without heating it. When the Lindeman melting condition on the magnitudes of atomic displacements (Sec. 1.3.) is met, the cold melting occurs. The acoustical frequency softening under EDTI (Sec. 4.1.) can make this process easier to occur.

Another question, which is not clear completely is the origin of the intensive point and other types defect formation under PLA. Moreover, laser-induced defects formation in surface layers begins, according to the experiments [183—185] long before the melting and cannot be explained in the frame of pure thermal model. Here, the certain role plays, probably the defect formation, induced by dense electron-hole plasma on the first stage of PLA [185, 186]. Recently, the EDT-model of the point defect formation was developed [285], which takes into account the joint effect of the three factors of laser action: the local electronic exitation, deformation which reduces the defect formation energy and the heating. The EDT theory of defect formation yilds the quantative results which are in agreement with experimental data on the point defect formation in Ge and GaAs [285].

3. In connection with the unsolved problems of PLA physics the primary importance gain the new methods of diagnostics of the ultrafast phase transformations on the semiconductor and metal surfaces. Aside of nonlinear optical methods, described in details in Sec.2, which use the sum and difference frequencies optical harmonics, generation on reflection, one must point out the pico- and femtosecond CARS spectroscopy of optical and acoustical phonons, and plasmons in laser exited crystalls. "The hibrid" technique of laser picosecond electrography [191, 192] was developed recently, which is very interesting and informative technique of investigation of structural transformations in subsurface layers of laser exited crystals in real time scale. In it the picosecond light pulses are used for generation (due to the external photoeffect from cathode surface) of powerful short synchronized in time with laser pulses electron bunches which after accelaration in pulse highvoltage electric field may be directed to the laser-exited region under investigation for the one instant diffraction picture registration and its evolution with picosecond resolution. The first data about the picosecond dynamics of laser-induced phase transformations in crystalline alluminium have been already obtained with the help of this technique [191].

The visual pictures of irradiated by powerful femtosecond annealing pulses surface regions of Si, taken sucsesivly with the interval of 100 fs with the use of stroboscopic technique were obtained recently in work by Douner et al [193].

Evidently this technique can be classified as ultrafast cinema with the interval between the cadres, which is determined by shining pulse duration — in this case 100 fs. The interesting data refering to first PLA phase dynamics in Si, were obtained, by Malvezzi et al by registration of the photo emission of charged particles (in picosecond diapason) [194]. All these new highly informative methods of laser diagnostics of fast phototransformation on the surfaces of solids are likely to diminish substantially the number of yet unsolved questions about the nature and deails of these processes.

4. In this book we repeatedly stressed the fact that speaking about a powerful laser exitation of subsurface layers of semiconductors and metals, one must not forget that the energy needed for this must be injected into the crystal through its surface. Under the action of laser light, as was described in details in Sec. 3.7. the interferential periodical structures of surface relief are formed, which can strongly deminish the reflection coefficient. According to the data and the results of Sec. 3.7.2 due to this surface absorbtivity is strongly enhanced. These and other type instabilities (for example DDI (Sec. 4)), which change the surface absorptivity, must be taken into acount while analysing all real experiments, related to laser exitation of subsurface layers of solids, including PLA experiments, laser drilling, welding and hardening of metals, laser action on biological objects and so on. The full account of these effect can be made on the basis of the nonlinear theory of the surface interferencial instabilities in developing of which only the first steps have been made (Sec. 3.6.1. — 3.6.5).

REFERENCES

1. J.Ready, Deistvie moshchnogo lasernogo izlucheniya (Action of Powerful Laser Radiation), Mir., M., 1974; Promyshlennoe primenenie laserov (Industrial Application of Lasers), Mir., M., 1981.

2. S.I.Anisimov, Ya.A.Imas, G.S.Romanov, and Yu.V.Khodyko, Deistvie izlucheniya bol'shoi moshchnosti na metally (Interaction of High-Power Radiation with Metals), Nauka, M., 1970.

3. W.Duley, Laser processing of Materials, Plenum Press, New York, 1983.

4. N.G.Basov and G.V.Sklizkov, Nagrev i szhatie termoyadernykh mishenei lazerom (Laser Heating and Compression of Thermonuclear Targets), VINITI, M., 1982.

5. I.B.Khaibullin, E.I.Shtyrkov, M.M.Zaripov et al., Rad. Eff. 36, 225 (1978).

6. L.V.Drurechenskii, G.A.Kachurin, N.V.Nidaev, and L.S.Smirnov, Impul'snyi otzhig poluprovodnikovykh materialov (Impulsive Annealing of Semiconductor Materials), Nauka, Novosibirsk, 1982.

7. S.D.Ferris, H.J.Leamy, and J.M.Poate (Eds.), Laser-Solid Interaction and Laser Processing, Amer. Inst. Phys., New York, 1979.

8. C.W.White and P.S.Peercy (Eds.), Laser and Electron Beam Processing of Electronic Materials, Academic Press, New York (1980).

9. C.L.Anderson, G.K.Cellar, and G.A.Rozgonyi (Eds.), Laser and Electron Beam Processing of Electronic Materials, ECS Inc., Princeton (1980).

10. J.F.Gibbon, L.D.Hess, and T.W.Sigmon (Eds.), Laser and Electron Beam Solid Interactions and Materials Processing, North-Holland, New York, 1981.

11. E.Kaldis (Eds.), Current Topics in Materials Science, North-Holland, Amsterdam, 1982, Vol. 8.

12. B.R.Appleton and G.K.Cellar (Eds.), Laser and Electron Beam Interaction with Solids, North-Holland, Amsterdam, 1982.

13. C.V.Shank, Science 219, 1027 (1983). R.L.Fork, B.I.Greene, and C.V.Shank, Appl. Phys. Lett. 38, 671 (1981). R.L.Fork, C.V.Shank, and R.Yen, ibid. 41, 273 (1982).

14. K.B.Eisenthal and R.Hochstrasser (Eds.), Picosecond Phenomena. III, Springer-Verlag, New York, 1982 (Springer Series in Chemical Physics, Vol. 23).

15. J.G.Fujimoto, A.M.Weiner, and E.P.Ippen, Appl. Phys. Lett. 44, 832 (1984).

16. C.V.Shank, R.Yen, and C.Hirlimann, Phys. Rev. Lett. 50, 454 (1983), 51, 900.

17. F.V.Bunkin, N.A.Kirichenko, and B.S.Luk'yanchuk, Usp. Fiz. Nauk 138, 45 (1982) (Sov. Phys. Usp. 25, 662 (1982); Izv. Akad. Nauk SSSR, Ser. Fiz. 48, 1485 (1984) (Bull. Acad. Sci. USSR. Phys. Ser.

18. S.A.Akhmanov, M.F.Galyautdinov, N.I.Koroteev et al., Kvant. Elektron. 10, 1977 (1983) (Sov. J. Quantum Electron. 13, 687; Opt. Commun. 43, 202 (1983).

19. H.W.K.Tom, T.F.Heinz, and Y.R.Shen, Phys. Rev. Lett. 51, 1983 (1983).

20. a) Abstracts of Reports at the 5th All-Union Conference on Nonresonant Interaction of Optical Radiation with Matter, Leningrad (1981);
b) Abstracts of Reports at the 6th All-Union Conference on Nonresonant Interaction of Optical Radiation with Matter, Leningrad (1984).

21. S.A.Akhmanov and N.I.Koroteev, Metody nelineinoi optiki v spektroskopii rasseyaniya sveta (Methods of Nonlinear Optics in the Light Scattering Spectroscopy), Nauka, M. (1981).

22. A.R.Beattie and P.T.Landsberg, Proc. R. Soc. London 249, 16 (1959).

23. A.Haug, Solid State Electron. 21, 1281 (1978).

24. E.J.Yoffia, Phys. Rev. B21, 2415 (1980).
25. L.V.Keldysh, Usp. Fiz. Nauk 100, 514 (1970) (Sov. Phys. Usp. 13, 291 (1970).
26. W.P.Dumke, Phys. Lett. A78, 477 (1980).
27. R.Biswas and V.Ambegaokar, Phys. Rev. B26, 1980 (1982).
28. C.L.Tang and D.J.Erskine, Phys. Rev. Lett. 51, 840 (1983).
29. E.Conwell and Vassell, IEEE Trans. ED-13, 22 (1966).
30. J.Shah, J. Phys. (Paris) 42, Suppl. N 10, C7-455, (1981).
31. T.Rice, J.Hansel, T.Phillips, and G.Thomas, Elektronno-dyrochnaya zhidkost' v poluprovodnikakh (Electron-Hole Liquid in Semiconductors), Mir, M. (1980).
32. I.Van Vechten, J. Phys. (Paris) 44, Supp. N 10, C5-11 (1983).
33. E.J.Yoffa, ibid. 41, Suppl. N 5, C4-7 (1981).
34   Phonon Scattering in Solids, Plenum Press, New York (1976).
35. D.von der Linde, J.Kuhl, and H.Klingenberg, Phys. Rev. Lett. 44, 1505 (1980).
36. R.G.Ulbrich, V.Narayanamurti, and M.A.Chin, ibid. 45, 1432.
37. M.Greenstein, M.A.Tamor, and J.P.Wolfe, Solid State Commun. 45, 355 (1983).
38. Northrop G.A. and Wolf J.P., Bull. Amer. Phys. Soc., 1983, 28, 252.
39. Kurz H., Lompre L.A., and Liu J.M., J. Phys. (Paris), 1983, 44, Suppl. N 10, C5-23.
40. Wood R.F. and Giles G.E., Phys. Rev., 1981, B23, 2923. Wood R.F., Kirkpatrick J.R., and Giles G.E., ibid. 23, 5555.
41. Lietoila A. and Gibbons J.F., in Ref. 10, p. 23.
42. Orbach R., IEEE Trans., 1967, SU-14, 140.
43. Rissel Kh. and Ruge I., Ionnaya implantatsiya (Ion Implantation), Nauka, M., 1983.
44. Wittmer W. and Rozgonyi G.A., in: Ref. 11, Ch. 1.
45. Bertolotti M. and Vitali G., ibid., Ch. 2.
46. Compaan A., Contreras G., Cardona M., and Axmann R., J. Phys. (Paris), 1983, 44, Supp. N 10, C5-197.
47. Wagner J., Compaan A., and Axmann R., ibid., Ct-61.
48. Slaoui A., Fogarassy E., Muller J.C., and Siffert P., ibid., Ct-65.
49. Alferov Zh.I., Abakumov V.N., Koval'chuk Yu.V., Ostrovskaya G.V., Portnoi E.L., Smirnitskii V.B., and Sokolov I.A., Fiz. Tekh. Poloprovodn., 1983, 17, 235, (Sov. Phys. Semicond., 1983, 17, 152.
50. Shtyrkov E.I., Khaibullin I.B., Galyautdinov N.F., and Zaripov M.M., Opt. Spektrosk., 1975, 38, 103.
51. Andreev A.V., and Akhmanov S.A., Preprint N 26/1981, Moscow, 1981; Andreev A.V., Akhmanov S.A., and Kov'ev E.K., Izv. Akad. Nauk SSSR, Ser, Fiz., 1983, 47, 1898.
52. Ziman J., Principles of the Theory of Solids, Cambridge University Press, 1964 (Russ. Transl. Mir., M., 1974).
53. Stritzler Q., Pospieszczyk A., and Tagle A., Phys. Rev. Lett., 2982, 47, 356.
54. Larson B.C., White C.W., Noggle T.S., and Mills D.M., Phys. Rev. Lett., 1982, 48, 337.
55. Larson B.C., White C.W., Noggle T.S., Barhorst J.F., and Mills D.M., in: Fef. 12, p. 13.
56. Williams R.T., Kablet M.N., Long J.P., Rife J.C., and Royt R.T., ibid., p. 97.
57. Thomson M.O., Calvin G.J., Mayer J.W., Hammond B.B., Paulter N., and Peercy P.S., ibid., p. 209.
58. White C.W., ibid., p. 109.
59. Van Vechten I.A., and Compaan A.A., Solid State Commun. 1981, 39, 867.
60. Bok J., J. Phys. (Paris), 1983, 44, Suppl. N 10, C5-3.

61. Compaan A., Lo H.W., Lee M.C., and Aydinli A., Phys. Rev., 1982, B26, 1079.
62. Von der Linde D., Wartmann G., and Ozols A., in: Laser-Solid Interactions and Transient Thermal Processing of Materials, edited by J.Narajan, W.L.Brown, and L.A.Lemons, Academic Press, New York, 1983.
63. Von der Linde D. and Wartmann G., Appl. Phys. Lett., 1982, 41, 700.
64. Wood R.F., Rasolt M., and Jellison G.E., in: Ref. 12, p. 61.
65. Compaan A., Lee M.C., Lo H.W., Trott G.J., and Aydinly A., J. Appl. Phys., 1983, 54, 5950.
66. Von der Linde D., Wartmann G., and Compaan A., Appl. Phys. Lett., 1983, 43, 613.
67. Van Vechten I.A., in: Ref. 12, p. 49.
68. Kaganov M.I., Lifshitz I.M., and Tanatarov L.V., Zh. Eksp. Teor. Fiz., 1956, 34, 232 (Sov. Phys. JETP, 1957, 4, 173); Anisimov S.I., Kapeliovich B.L., and Perel'man T.L., Zh. Eksp. Teor. Fiz., 1974, 66, 776 (Sov. Phys. JETP, 1974, 39, 375).
69. Easley G.L., Phys. Rev. Lett., 1983, 51, 2140; 1987, V. 58, p. 1680.
70. Kapaev V.V., Kopaev Yu.V., and Molotkov S.N., Mikroelektronika, 1983, 12, 499.
71. Suslov I.M., Pis'ma Zh. Eksp. Teor. Fiz., 1984, 39, 449 (JETP Lett., 1984, 39, 544).
72. Fujimoto J.G., Liu J.M., Ippen E.P., and Bloembergen N., in: Ref. 75, p. 140.
73. Fujimoto J.G., Shevel S.G., and Ippen E.P., Solid State Commun., 1984, 49, 605.
74. Piskarskas A.S. et al., Parametricheskie generatory sveta i pikosekundnaya spektroskopiya (Parametric Light Generations and Picosecond Sectroscopy), edited A.S. Piskarskas, Mokslas, Vilnyus (1983).
75. Auston D.H. and Eisnthal K.B. (Eds.), Ultrafast Phenomena IV, Springer-Verlag, New York, 1984 (Springer Series in Chemical Physics, Vol. 38).
76. Shapiro S. (Ed.), Sverkhkorotkie svetovye impul'sy (Ultrashort Light Pulses), Mir. M. (1981).
77. Shank C.V., Ippen E., and Shapiro S. (Eds.), Picosecond Phenomena, Springer-Verlag, New York (1978) (Springer Series in Chemical Physics, Vol. 4).
78. Sverkhbystrye protsessy v spektroskopii (Ultrafast Processes in Spectroscopy), Valgus, Tallin (1979).
79. Shank C.V., Hochstrasser R., and Kaiser W. (Eds.), Picosecond Phenomena, II, Springer-Verlag, New York (1980) (Springer Series in Chemical Physics, Vol. 14).
80. "Picosecond light phenomena" in: Abstracts of Reports at the 10th, 11th, and 12th All-Union Conferences on Coherent and Nonlinear Optics, Kiev (1980); Erevan (1982); Moscow (1985).
81. Auston D.H., Surko C.M., Venkatesan T.N.C., Slusher R.E., and Solovchenko A., Appl. Phys. Lett., 1978, 33, 437.
82. Auston D.H., Solovchenko J.A., Simons A.L., Surko C.M., and Venkatesan C., ibid., 1979, 34, 777.
83. Von der Linde D. and Fabricius N., ibid., 1982, 41, 991.
84. Lui J.M., Kurz H., and Bloembergen N., ibid., 1982, 41, 643.
85. Combescot M., Phys. Rev. Lett., 1983, 51, 519.
86. Lui M., Yen R., Kurz H., and Bloembergen N., Appl. Phys. Lett., 1981, 39, 755.
87. Bloembergen N., Kurz H., Lui J.M., and Yen R., in: Ref. 12, p. 37.
88. Yen R., Lui J.M., Kurz H., and Bloembergen N., Appl. Phys., 1982, A27, 153.
89. Lompre L.A., Kui J.M., and Bloembergen N., Appl. Phys. Lett., 1984, 44, 3.
90. Van Driel H.M., Lompre L.A., and Bloembergen N., ibid., 285.
91. Aydinli A., Lo H.W., and Compaan A., Phys. Rev. Lett., 1981, 46, 1640.
92. Compaan A., Aydinli A., Lee M.C., and Lo H.W., in: Ref. 12, p. 43.

192

93.   Moss S.C. and Marquardt Q., ibid., p. 79.
94.   Lowndess D.H, Phys. Rev. Lett., 1982, 48, 267.
95.   Lo H.W. and Compaan A., Phys. Rev. Lett., 1980, 44, 1604. Compaan A., Lo W.H., Aydinli A., and Lee M.C., in: Ref. 10, p. 15.
96.   Lee M.C., Lo H.W., Aydinli A., and Compaan A., Appl. Phys. Lett., 1981, 38, 499.
97.   Wood R.F., Lowndess D.H., and Giles G.E., in: Ref. 12, p. 67.
98.   Schmidt M., and Dransfeld K., Appl. Phys., 1982, A28, 211.
99.   Emel'yanov V.I. and Uvarova I.F., Akust. Zh., 1985, 31, 481 (Sov. Phys. Acoust., 1985, 31, 284).
100. Portnoi E.L., Koval'chuk Yu.V., Ostrovskaya G.V., Piskarskas A.S., Skopina V.I., Smil'gyavichyus V.I., and Smirnitski V.V., Pis'ma Zh. Tekh. Fiz., 1982, 8, 462 (Sov. Tech. Phys. Lett., 1982, 1, 201).
101. Abakumov V.N., Alferov Zh.I., Koval'chuk Yu.V., and Portnoi E.L., ibid., 1982, 8, 966 (Sov. Tech. Phys. Lett., 1982, 8, 417).
102. Alferov Zh.I., Koval'chuk Yu.V., Smol'skii O.V., and Sokolov I.A., ibid., 1983, 9, 897 (Sov. Tech. Phys. Lett., 1983, 9, 387).
103. Alferov Zh.I., Koval'chuk Yu.V., Pogorel'skii Yu.V., Smol'skii O.V., and Sokolov I.A., ibid., 1373 (Sov. Tech. Phys. Lett., 1983, 9, 591.
104. Aksenov V.P. and Zhurkin B.G., Dokl. Akad. Nauk SSSR, 1982, 265, 1365. (Sov. Phys. Dokl., 1982, 27, 630).
105. Bloembergen N. and Pershan P., Phys. Rev., 1962, 128, 606.
106. De Martini F. and Shen Y.R., Phys. Rev. Lett., 1976, 36, 216.
107. De Martini F., Colocci M., Kohn S.E., and Shen Y.R., ibid., 1977, 38, 1223.
108. Shen Y.R., Chen C.K., and Castro A.R.B., in: Proceedings of the VIII International Conference on Raman Spectroscopy, Ottawa, Canada, 1980, edited by W.F.Murphy, North-Holland, New York, 1980, p. 659.
109. Chen C.K., Heinz T.F., Ricard D., and Shen Y.R., Phys. Rev. Lett., 1981, 46, 1010.
110. Heinz T.F., Chen C.K., Ricard D., and Shen Y.R., Chem. Phys. Lett., 1981, 83, 180.
111. Aktsipetrov O.A. and Mishina E.A., Pis'ma Zh. Eksp. Teor. Fiz., 1983, 38, 442 (JETP Lett., 1983, 38, 535).
112. Heritage J.P. and Bergman J.G., in: Light Scattering in Solids, edited by J.L.Birman, H.Z.Cummins, and K.K.Rebane, Plenum Press, New York, 1979, p. 167.
113. Akhmanov S.A., Galyautdinov M.F., Koroteev N.I., Paityan G.A., Khaibullin I.B., Shtyrkov E.I., and Shumai L., Pis'ma Zh. Tekh. Fiz., 1984, 10, 1900 (sic); Akhmanov S.A., Galiautdinov M.F., Koroteev N.I., Paitian G.A., Khaibullin I.B., Shtykov E.I., and Shumay I.L., J. Opt. Soc. Amer., 1985, B2, 283.
114. Akhmanov S.A., Galyautdinov M.F., Koroteev N.I., Paityan G.A., Shumai I.L., et al., Izv. Akad. Nauk SSSR, Ser. Fiz., 1985, 49, 506.
115. Cheng H. and Miller P.B., Phys. Rev., 1964, A134, 683.
116. Bloembergen N., Chang R.K., Jha S.S., and Lee C.H., ibid., 1968, 174, 813.
117. Akhmanov S.A., Khokhlov R.V., Problemy nelineinoi optiki (Problems in Nonlinear Optics), VINITI, M., 1964.
118. Bloembergen N., Nonlinear Optics, Benjamin, New York, 1965 (Russ. Transl. Mir., M., 1966).
119. Sing S., in: Spravochnik polazeram (Laser Handbook), edited by A.M.Prophorov, Sov. radio, Moscow, 1978, Vol. 2, p. 237.
120. Meisner L.B. and Saltiel S.M., ibid., p. 271.

121. Nye J.F., Physical Properties of Crystals, Clarendon Press, Oxford, 1957 (Russ. Transl. Mir., M., 1967).

122. Ducuing J. and Flytzanis C., in: Optical Properties of Solids, edited by F. Abeles, North-Holland, Amsterdam, 1972, p. 863.

123. Ducuing J. and Bloembergen N., Phys. Rev. Lett., 1963, 10, 474.

124. Lee C.H., Chang R.K., and Bloembergen N., ibid., 1967, 18, 167.

125. Aktsipetrov O.A. and Mishina E.D., Dokl. Akad. Nauk SSSR, 1984, 274, 62 (Sov. Phys. Dokl., 1984, 29, 37).

126. Shank C.V., Yen R., Hirlimann C. Phys. Rev. Lett, 1983, V. 51, p. 900.

127. Malvezzi A.M., Lui J.M., and Bloembergen N., Appl. Phys. Lett., 1984, 45, 1019.

128. Diels J.C., Laser Focus, 1983, 19, 26.

129. Guidotti D., Driscoll T.A., and Gerritsen H.J., Solid State Commun, 1983, 46, 337.

130. Driscoll T.A. and Guidotti D., Phys. Rev., 1983, B28, 1171.

131. Aspnes D.E. and Studna A.A., ibid., 27, 985.

132. Belonozhko A.B., Emel'yanov V.I., Paityan G.A., and Sumbatov A.A., Vestm. Mosk. un-ta, Ser. Fizika, Astronomiya, 1984, 26, N 4, 67.

133. Born M. and Wolf E., Principles of Optics, Pergamon Press, Oxford, 1970 (Russ. Transl. Nauka M., 1973).

134. Lutwin J., Moss D.J., Sipe J.E., and van Driel H.M., "Comparison of nanosecond and picosecond harmonic generation from centrosymmetric semiconductors", Post-Deadline Paper WII29-1, presented at the 13th International Quantum Electronic Conference, Anaheim, USA, 1984.

135. Birnbaum M.J., Appl. Phys., 1965, 36, 3688.

136. Bonch-Bruevich A.M., Kochengina M.K., Libenson M.N., Pudkov S.D., and Trubaev V.V., Izv. Akad. Nauk SSSR, Ser. Fiz., 1982, 46, 1186.

137. Prophorov A.M., Sychugov V.A., Tishchenko A.V., and Khakimov A.A., Pis'ma Zh. Tekh. Fiz., 1982, 8, 961, 1409 (Sov. Tech. Phys. Lett., 1982, 8, 415, 605).

138. Gromov G.G. and Ufimtsev V.B., Dokl. Akad. Nauk SSSR, 1983, 272, 1405; Pis'ma Zh. Tekh. Fiz., 1983, 9, 580 (Sov. Tech. Phys. Lett., 1983, 9, 249).

139. Young J.F., Preston J.S., van Driel H.M., and Sipe J.E., Phys. Rev., 1983, B27, 1141, 1155.

140. Oron M. and Sorensen G., Appl. Phys. Lett., 1979, 35, 782.

141. Fauchet P.M. and Siegman A.E., ibid., 1982, 40, 824; in Ref. 75, p. 129; Appl. Phys., 1983, A32, 135.

142. Bazakutsa P.V., Prophorov A.M., Sychugov V.A., and Tishchenko A.V., Pis'ma Zh. Tekh. Fis., 1983, 9, 705 (Sov. Tech. Phys. Lett., 1983, 9, 303).

143. Bazhenov V.V., Bonch-Bruevich A.M., Libenson M.N., Makin V.S., Pudkov S.D., and Trubaev V.V., Pis'ma Zh. Tekh. Fiz., 1983, 9, 932 (Sov. Tech. Phys. Lett., 1983, 9, 402).

144. Isenor N.R., Appl. Phys. Lett., 1977, 31, 148.

145. Jain A.K., Kulkarni V.N., Sood D.K., and Uppal J.C., J. Appl. Phys., 1981, 52, 4882.

146. Anisimov V.N., Baranov V.Yu., Bol'shov L.A., Dykhne A.M., Malyuta D.D., Pis'men-nyi V.D., Sbrant A.Yu., and Stepanova M.A., Poverkhnost, 1983, N 7, 138.

147. Temple P.A. and Soleau M.J., IEEE J., 1981, QE-17,

148. Konov V.I., Prokhorov V.A., Sychugov V.A., Tishchenko A.V., and Tokarev V.N., Zh. Tekn. Fiz., 1983, 53, 2283 (Sov. Phys. Tech. Phys., 1983, 28, 1404).

149. Keilmann F. and Bai Y.H., Appl. Phys., 1982, A29, 9.

150. Koo J.C. and Slusher R.E., Appl. Phys. Lett., 1976, 28, 614.

151. Brueck S.R.J. and Ehrich D.J., Phys. Rev. Lett., 1982, 48, 1678.

152. Keilmann F., ibid., 1983, 51, 2097.

153. Young J.F., Sipe J.E., and van Driel H.M., Optics Lett., 1983, 8, 431.

154. Heinz T.F., Loy M.M., and Thompson W.A., Phys. Rev. Lett., 1985, 54, 63.

155. Viktorov I.A., Zvukovye poverkhnostnye volny v tverdykh telakh (Surface Acoustic Waves in Solids), Nauka, M., 1981.

156. Sandercock R., Solid State Commun, 1978, 26, 547.

157. Rowell N.L. and Siegman G.I., Phys. Rev., 1978, B18, 2598.

158. Landau L.D. and Lifshitz E.M., Mekhanika sploshnykh sred; Teoria uprugosti, Fizmatgiz, M., 1953 (Engl. Transl. Fluid Mechanics; Theory of Elasticity, Pergamon Press, Oxford, 1959).

159. Jha S.S., Kirtley J.R., and Tsang J.C., Phys. Rev., 1980, B22, 3973.

160. Emel'yanov V.I., Zemskov E.M., and Seminogov V.N., Poverkhnost', 1984, 2, 38.

161. Emel'yanov V.I. and Koroteev N.I., Usp. Fiz. Nauk, 1981, 135, 345 (Sov. Phys. Usp., 1981, 24, 864).

162. Zel'dovich B.Ya. and Sobel'man I.I., Usp. Fiz. Nauk, 1970, 101, 3 (Sov. Phys. Usp., 1970, 13, 307).

163. Kats A.V. and Maslov V.V., Zh. Eksp. Teor. Fiz., 1972, 62, 496 (Sov. Phys. IETP, 1972, 35, 258).

164. Bunkin F.V., Samokhin A.A., and Fedorov N.V., Zh. Eksp. Teor. Fiz., 1969, 56, 1057 (Sov. Phys. JETP, 1969, 29, 568).

165. Emel'yanov V.I. and Seminogov V.N., Zh. Eksp. Teor. Fiz., 1984, 86, 1026 (Sov. Phys. JETP, 1984, 59, 598).

166. Maslennikov V.L., Prokhorov A.M., Sychugov V.A., and Tishchenko A.V., Pis'ma Zh. Tekh. Fiz., 1983, 9, 679 (Sov. Tech. Phys. Lett., 1983, 9, 292).

167. Fiziko-khimicheskie svoistva elementov (Physico-Chemical Properties of the Elements), Naukova dumka, Kiev, 1965.

168. Emel'yanov V.I. and Seminogov V.N., Kvant. Elektron., 1984, 11, 871 (Sov. J. Quantum Electron., 1984, 11, 591).

169. Golubenko G.A., Samokhin A.A., and Sychugov V.A., ibid., 1850 (Sov. J. Quantum Electron., 1984, 11, 1239).

170. Tribel'skii M.I. and Gol'dberg S.M., Pis'ma Zh. Tekh. Fiz., 1982, 8, 1227.

171. Emel'yanov V.I., Zemskov E.M., and Seminogov V.N., Kvant. Elektron., 1983, 10, 2389 (Sov. J. Quantum Electron., 1983, 10, 1556).

172. Emel'yanov V.I., Zemskov E.M., and Seminogov V.N., ibid., 1984, 11, 2283 (Sov. J. Quantum Electron., 1984, 11, 1484).

173. Bonch-Bruevich A.M., Libenson M.N., and Makin V.S., Pis'ma Zh. Tekh. Fiz., 1984, 10, 3 (Sov. Tech. Phys. Lett., 1984, 10, 1).

174. Samokhin A.A., Kvant. Elektron., 1983, 10, 2022 (Sov. J. Quantum Electron., 1983, 10, 1347); Preprint N 34, PLAN USSR, Moscow, 1984.

175. Fiquerira J.F. and Thomas S.J., in: Surface Studies with Lasers, edited by F.R. Ausenegg, A. Leitner, and M.E. Lippitsch, Springer-Verlag, New York (1982), p. 212 (Chemical Physics, Vol. 33).

176. Landau L.D. and Lifshitz, Statisticheskaya fizika, Nauka, M., 1976, Pt. 1, p. 281. IEngl. Transl. Statistical Physics, Pergamon Press, Oxford, 1980).

177. Anisimov S.I., Tribel'skii M.I., and Epel'baum Ya.G., Zh. Tekh. Fiz., 1980, 78, 1597 (sic).

178. Prokhorov A.M., Sychugov V.A., Tishchenko A.V., and Khakimov A.A., Pis'ma Zh. Tekh. Fiz., 1983, 9, 65 (Sov. Tech. Phys. Lett., 1983, 9, 28).

179. Bazhenov V.V., Bonch-Bruevich A.M., Libenson M.N., Makin V.S., Pudrov S.D., and Trubaev V.V., ibid., 1268 (Sov. Tech. Phys. Lett., 1983, 9, 544).

180. Aksenov V.P. and Zhurkin B.G., Preprint N 56, 194, FIAN, Moscow (1982).

181. Regel' A.R. and Glazov V.M., Fizicheskie svoistva elektronnykh rasplavov (Physical Properties of Electronic Melts), Nauka, M., 1980.

182. Kikoin I.K. (Ed.), Tablitsy fizicheskikh velichin, Spravochnik (Tables of Physical Quantities Handbook), Atomizdat, M., 1976.

183. Kashkarov P.K., Kiselev V.F., and Petrov A.V., Poverkhnost', 1982, 12, N 10, 47.

184. Chabal Y.J., Rowe J.E., and Christman S.B., J. Vacuum Sci. Technol., 1982, 20, 763.

185. Karjagin S.N., Kashkarov P.K., Kiselev V.F., and Petrov A.V., Surf. Sci., (Letters), 1984, 146, L582.

186. Wautelet M., Failley-Lovato M., and Laude L., J. Phys., 1980, C13, 5504.

187. Emel'yanov V.I. and Seminigov V.N., Poverkhnost, 1985, N 11, 142.

188. Keilmann F., in: Ref. 175, p. 207.

189. Jacoboni C., Canali C., Ottaviani G., and Albergi A., Solid State Electron., 1977, 20, 77.

190. Emel'yanov V.I., "Nonlinear-optical deformation of the acoustic sybsystem and ultrafast melting of semiconductor surfaces by powerful short laser pulses", Moscow State University, Physics Fucalty, Preprint N 5, Moscow (1985).

191. Mourou G. and Williamson S., Appl. Phys. Lett., 1982, 41, 44.

192. Akhmanov S.A., Bagratashvili V.N., Golubkov V.V., Zgurskii A.V., Ishcheno A.A., Krikunov S.A., and Tunkin V.G., Pis'ma Zh. Tekh. Fiz., 1985, 11, 162 (Presented at the 3rd International Conference on Ultrafast Processes) (sic).

193. Downer M., Fork R.L., and Shank C.V., in: Ref. 75, p. 114.

194. Malvezzi A.M., Kurz H., and Bloembergen N., ibid, p. 118.

195. Eisenmenger W. (Ed.), Phonon Scattering in Condensed Matter, Springer-Verlag, New York (1984).

196. Wolf S.A., Cubbertson J.C., Strom U., and Klein P.B., Phys. Rev., 1984, B29, 7054.

197. Avanesyan S.M., Akhmanov S.A., Bonch-Osmolovskii M.M., Gusev V.E., Zheludev N.I., and Petrosyan E.G., in: Ref. 20, p. 148.

198. Avanesyan S.A., Gusev V.E., and Zheludev N.I., Appl. Phys., 1985, A29, 5.

199. Gnosheng Z., Fauchet P.M., and Siegman A.E., Phys. Rev., 1982, B26, 5366.

200. Wood R.W., Philos. Mag., 1902, 4(6), 396; Phys. Rev., 1935, 48, 928.

201. Rayleigh, Philos. Mag., 1907, 14(6), 60.

202. Young J.F., Sipe J.E., and van Driel H.M., Phys. Rev., 1984, B30, 2002.

203. Soileau M.J., IEEE J., 1984, QE-20, 464.

204. Bonch-Bruevich A.M., Libenson M.N., and Makin V.S., in: Fef., 20b, p.4.

205. Reinish R. and Neviere M., Opt. Engng., 1981, 20, 629.

206. Andreev A.V., Usp. Fiz. Nauk, 1985, 145, 113 (Sov. Phys. Usp., 1985, 28, 70).

207. Bazakutsa P.V., Sychugov V.A., and Prokhorov A.M., Kvant. Elektron. (Moscow), 1984, 11, 2127 (Sov. J. Quantum Electron., 1984, 11, 1420).

208. Ordal M.A., Long L.L., Bell R.J., Bell S.E., Alexander R.W., and Ward C.A., Appl. Optics, 1983, 22, 1099.

209. Avrutskii I.A., Bazakutsa P.V., Sychugov V.A., and Tishchenko A.V., Pis'ma Zh. Tekh. Fiz., 1984, 10, 1333 (Sov. Tech. Phys. Lett., 1984, 10, 563.

210. Ursu I., Mihailescu I.N., Popa Al., Prokhorov A.M., Konov V.I., Ageev V.P., and Tokarev V.N., Appl. Phys. Lett., 1984, 45, 365.
211. Konov V.I. and Tokarev V.N., "Size effects accompanying laser irradiation of periodic surface structures", (In Russian) Preprint N 70, Institute of Optical Physics of the USSR Academy of Sciences, Moscow (1985).
212. Wartman G., Kemmler M., and von der Linde D., Phys. Rev., 1984, B30, 4850.
213. Emel'yanov V.I. and Seminogov V.N., in: Abstacts of Reports at the 12th All-Union Conference on Coherent and Nonlinear Optics, Moscow, 1985, p. 399.
214. Akhmanov S.A., Seminogov V.N., Sokolov V.I., JETP, 1987, v. 94, p. 1654.
215. Gorodetzky G., Kanicki J., Kazyaka T., and Melcher R., Appl. Phys. Lett., 1985, 40, 547.
216. Makshantsev B.I. and Pilipetski N.F., Appl. Phys., 1985, A36, 205.
217. Compaan A., Lee M.C., Trott G.J., Phys. Rev. B, 1985, v. 32, N 10, pp. 6731–6741.
218. Guidotte D., Driscoll T.A., Nuovo Cim., 1986, 8D, 385–416.
219. Heinz T.F., Loy M.M.T., Thompson W.A., J. Vac. Sci. Technol. B., 1985, 3, N 5, 1467–1470.
220. Akhmanov S.A., Emel'yanov V.I., Koroteev N.I., Seminogov V.N., Sov. Phys. Usp., 1985, 147, 675–685.
221. Guyout-Sionnest P., Chen W., Shen Y.R., Phys. Rev. B, 1986, 33, N 12, 8254–8263.
222. Guyot-Sinnest P., Shen Y.R., Phys. Rev. B, 1987, 35, N 11, 4420–4426.
223. Aktsipetrov O.A., Baranova I.M., Kuljuk L.L., Petukhov A.V., Arushanov E.K., Shtanov A.A., Shutov D.A., Fiz. Tverd. Tela, 1986, 28, N 10, 3228–3230.
224. Wang C.C., Phys. Rev., 1969, 178, 1457.
225. Aktsipetrov O.A., Baranova I.M., Il'inskii Yu.A., Zh. Exp. Teor. Fiz., 1986, 91, 287.
226. Loy M.M.T., Heinz T.F., Marowsky G., Lectures, delivered at the VIII-th International Shool on Coherent and Nonlinear Optics, Tbilisi, USSR, 1987.
227. Shen Y.R., Ann. Rev. Mat. Sci., 1986, 16, 69–86.
228. Chen C.K., de Castro A.R.B., Shen Y.R., Phys. Rev. Lett., 1981, 46, 145.
229. Heinz T.F., Chen C.K., Ricard D., Shen Y.R., Phys. Rev. Lett., 1982, 48, 478.
230. Heinz T.F., Tom H.W.K., Shen Y.R., Phys. Rev. A, 1983, A28, 1883.
231. Boyd G.T., Shen Y.R., Hänsch T.W., "CW SHG as a surface microprobe"., Opt. Lett., 1986, 11, 97.
232. Berkovic G., Rasing Th., Shen Y.R., J. Chem. Phys., 1986, 85, N 12, 7374–7376.
233. Guyot-Sionnest P., Hsiung H., Shen Y.R., Phys. Rev. Lett., 1986, 57, N 23, 2963–2966.
234. Zhu X.D., Suhr H., Shen Y.R., Phys. Rev.B, 1987, 35, N 6, 3047–3050.
235. Rasing Th., Berkovic G., Shen Y.R., Grubb S.G., Kim M.W., Chem. Phys. Lett., 1986, 130, N 1,2, 1–5.
236. Dick B., Gierulski A., Morowsky G., Reider G.A., Appl. Phys. B, 1985, B38, 107–116.
237. Aktsipetrov O.A., Akhmediev N.N., Mishina E.D., Novak V.R., Pis'ma v Zh. Exp. Teor. Fiz., 1983, 37, 175.
238. Hayden L.M., Kowel S.T., Srinivasan M.P., Opt. Commun., 1986.
239. Kemnits K., Bhattacharyya K., Hicks J.M., Pinto G.R., Eisenthal K.B., Heinz T.F., Chem. Phys. Lett., 1986, 131, N 4–5, 285–290.
240. Boyd G.T., Rasing Th., Leite J.R.R., Shen Y.R., Phys. Rev.B, 1984, B30, 510.
241. Aktsipetrov O.A., Baranova I.M., Dubinina E.M., Elovikov S.S., Elyutin P.V., Esikov D.A., Nikulin A.A., Fominykh N.N., Phys. Lett., 1986, A117, N 5, 239–242.

242. Belyi M.U., Robur L.I., Sushko A.M., Shaikevich I.A., Poverkhnost, Fiz., Mekhan., 1986, N 8, 54–57.
243. Wood R.F., Phys. Rev.B, 1986, 34, N 4, 2407–2415.
244. El-Adawi M.K., J. Appl. Phys., 1986, 60, N 7, 2256, ibid., p. 2260.
245. Larson B.C., White C.W., Noggle T.S., Mills D., Phys. Rev. Lett., 1982, 48, N 5, 337–340.
246. Gerritsen H.C., van Brug H., Bijkerk F., Murakami K., van der Wiel M.J., J. Appl. Phys., 1986, 60, N 5, 1774.
247. Wang C.C., Bomback J., Donlon W.T., Huo C.R., James J.V., Phys. Rev. Lett., 1986, 57, N 13, 1647–1650.
248. Moss D.J., Litwin J.A., van Driel H.M., Sipe J.E., Appl. Phys. Lett., 1986, 48, 1150.
249. Govorkov S.V., Koroteev N.I., Shumay I.L., Vestnik Moskovskogo Univ., Ser. Fiz., Astron., 1986, 27, N 3, 75.
250. Galjautdinov M.F., Govorkov S.V., Koroteev N.I., Khaibullin I.B., Shumay I.L., Shtyrkov E.I., ibid., 1986, 27, N 4, 99.
251. Gibbons J.F., Proc. IEEE, 1972, 60, 1062.
252. Govorkov S.V., Koroteev N.I., Shumay I.L., Izv. Akad. Nauk SSSR, ser. Fiz., 1986, 50, 683.
253. Tom H.W.K., Paper presented at VII-th IQEC/CLEO Conference, Baltimore, May 1987, USA.
254. Abdullaev A.Yu., Govorkov S.V., Koroteev N.I., Kashkarov P.K., Petrov G.I., Shumay I.L., Fiz. Tverd. Tela, 1987, 29, 1998.
255. Emel'yanov V.I., Koroteev N.I., Yakovlev V.V., Opt. i Spektroskopiya, 1987, 62, 1188.
256. Govorkov S.V., Zadkov V.N., Koroteev N.I., Shumay I.L., Pis'ma v Zh. Exp. Fiz., 1986, 44, 98. Solid State Commun., 1987, 62, 331.
257. Balkanski M., Wallis R.F., Haro E., Phys. Rev., 1983, B28, 1928.
258. Anastassakis E., Pinczuk A., Burstein E., Pollack E.H., Cardona M., Solid State Commun., 1970, 8, 133.
259. Papers by K.Kash, J.Shah, D.Block, A.Gossard, W. Weigmann, A.F. Levy, J. Hayes, P. Platzman, J. Tsang, J. Kash, S.S. Jha, R. Fork, et al., in: Physica, 1985, 134B, Amsterdam: North-Holl. Publ. Co.
260. Gusakov G.M., Komarnitskii A.A., Pis'ma v Zh. Tekhn. Fiz., 1987, 13, N 3, 166–170.
261. Mc Gilp J.F., Jeh J., Solid State Commun., 1986, 59, N 2, 91–94.
262. Wolff P.A., Juen S.J., Thomas G.A., Solid State Commun., 1986, 60, N 8, 645–648.
263. Weber M.G., Phys. Rev. B, 1986, 33, N 40, 6775–6779.
264. Biwer B.M., Pellin M.J., Shauer M.W., Guen D.M., Surf. Sci., 1986, 176, N 1–2, 377–396.
265. Emel'janov V.I., Seminogov V.N., Sokolov V.I. Kvant. Electr., 1987, v. 14, p. 33 (Sov. J. Quant. Electron.).
266. Emel'janov V.I., Seminogov V.N. The nonlinear regime of the laser generation of the capillar waves and the formation of the ordered surface structures preprint of the scientifical research centre on technological lasers of the Academy of Science of USSR. Troisk, N 15, 1986.
267. Emel'janov V.I., Konov V.N., Seminogov V.N., Tokarev V.N. The formation of the surface periodical structures under the action of the pulsed radiation of CO_2-laser on fused silica. Preprint of the General Physics Institute of the Academy of Science or USSR, N 89, Moscow, 1987 (J. Opt. Soc. Amer., B in press).

268. Prokhorov A.M., Svakhin A.C., Sichugov V.A., Tichenko A.V., Hakimov A.A., Kvant. Electron., 1983, v. 10, p. 906 (Sov. J. Quant. Electron.).

269. Konov V.I., Prokhorov A.M., Sichugov V.A., Tokarev V.N., Chapliev N.I. Preprint of Physical Institute of Academy of Science of USSR, N 122, Moscow, 1983, p. 26.

270. Bonch-Bruevich A.M., Imas Ya.A., Romanov G.S., Livenson M.N., Maltsev L.N. J. Zh. Tekh. Fiz., 1968, v. 38, p. 851.

271. Zavecs T.E., Safi M.A., Notice M. – Appl. Phys. Lett., 1975, v. 26, p. 165.

272. Walter C.T., Clauer A.H. – Appl. Phys. Lett., 1978, v. 33, p. 713.

273. Roessler D.M., Gregson V.G. – J. Appl. Optics, 1978, v. 17, p. 992.

274. Grinberg S.A., Emeljanov V.I., Sumbatov A.A. Proc. of the VII All-Union Conference on Crystal Growth, v. 1, p. 228, Moscow, 1988.

275. Emel'janov V.I., Seminogov V.N. – Kvant. Electron., 1987, v. 14, p. 47.

276. Abdullaev A.Yu., Govorkov S.V., Koroteev N.I., Petrov G.I., Shumay I.L. – Pis'ma Zh. Tekh. Fiz., 1986, v. 12, p. 1363.

277. Kanemitsu Y., Kuroda H., Nakada I. – Jap. J. Appl. Phys., 1986, v. 25, p. 1377.

278. Alferov Zh.I., Koval'chuk Y.V., Pogorel'skii Yu.V., Smol'skii O.V., Sokolof I.A. – Pis'ma Zh. Tekh. Fiz., 1983, v. 9, p. 1298.

279. Kopaev Yu.V. – Zh. Tech. Phys., 1985, v. 55, p. 2244.

280. Bugaev A.A., Klochkov A.V. – Fiz. Tverdogo Tela, 1984, v. 26, p. 3487.

281. Baskin B.L., Polyakov A.A., Trukhin V.N., Yaroshetzky I.D. – Pis'ma Zh. Tech. Phys., 1985, v. 11, p. 1251.

282. Merkle K.L., Vebbing R.H., Baumgart H., Phillip F. – in Ref. 12, p. 337.

283. Emel'yanov V.I., Uvarova I.F. – Izvestia Acad. Sci, USSR, Ser. Phys., 1986, v. 50, p. 1214.

284. Jost D., Lüthy W., Weber H., Sabathe R. – Appl. Phys. Lett., 1986, v. 49, p. 625.

285. Emel'janov V.I., Kashkarov P.K. The electron-deformational-thermal mechanism of the defect generation in the semiconductors under the action of the pulsed laser radiation. Preprint of the Physical faculty of Moscow State University, N 16, 1987.

286. Cottrel A.H. The theory of crystal dislocation, Blackie and Son LTD, London, 1964.

287. Veiko V.P., Dorofeev I.A., Imas Ya.I., Kalugina T.I., Livenson M.N., Shandibina G.D., Yakovlev E.B. – Pis'ma Zh. Tech. Phys., 1984, v. 10, p. 15; Izvestia Acad. Nauk USSR, ser. Phys., v. 49, p. 1236.

288. Pilecki S. – Archives of Mechanics, 1977, v. 29, p. 505.

289. Al'shitz V.I., Indenbom V.L. – in "The Dynamics of Dislocations", "Naukova Dumka", Kiev, 1975, p. 232.

290. Emeljanov V.I. The instabilities and the phase transitions under the action of the external energy pump. Proc. IV Intern. Symp. on selected problems of Statist. Mech. Dubna, USSR, 1988, p. 119.

291. Palatnik L.S., Cheremskoi P.G., Fuks M.Ya. Pori v pljonkakh (voids in the films), Moscow, Energoizdat, 1982.

292. Tom H.W.K., Aumiller G.D., Britto-Cruz C.H. Phys. Rev. Lett., 1988, v. 60, p. 1438.

293. Bagratashvili V.N., Banishev A.F., Gnedoi S.A., Emel'yanov V.I., Jerikhin A.N., Merzl'yakov K.S., Panchenko V.Ya., Seminogov V.N. The formation of the ring structures of the relief and the voids under the laser deposition of the metalic films. – Preprint of the Scientifical Research Centre on Technological Lasers of the Academy of USSR, N 32, Troisk, 1988.

294. Kim D.K., Shah R.P., Von der Linde. – in Ref. 12, p. 85.

295. Heinig K.H., Giller H.D. – Phys. Stat. Sol.(a), 1985, v. 92, p. 421.

296. Aleksandrov L.N. Kinetica kristallizatsii and perecristallizatsii poluprovodnikovikh plenok. (The kinetic of the crystallization and recrystallization of the semiconductor films). Novosibirsk, Nauka, 1985.

297. Borgoin J.C., Asomosa R. — J. de Phys. Coll , 5, Suppl. 10, 1983, v. 44, p. 175.

298. Germain P.J., Paesler M.A., Zellama K. In "Cohesion Properties of Semiconductors under Laser Irradiation", ed. D.Lande, 1983, p. 506.

299. Gsari C., Nihoul C., Marfaing J., Marine W. — Surface Sci., 1985, v. 162, p. 724.

300. Emel'yanov V.I., Seminogov V.N., Sokolov V.I. — Kvant. Electr., 1987, v. 14, p. 2028.

301. Geller G.K., Modification of silicon properties with lasers, electron beams and incoherent light. CRC Critical Reviews in Solid State and Materials Science, 1983, v. 12, N 3, p. 193.

302. Murakami K., Masuda K. Dynamic behaviour of picosecond and nanosecond pulsed laser annealing in ion-implanted semiconductor. In: Semiconductors Probed by Ultrafast Laser Spectroscopy, v. 11, 1984, Academic Press, p. 171.

303. Van Driel H.M. — Physics of Pulsed Laser Processing of Semiconductor. In: Ref. 302, p. 57.

304. Van Vechten J.A. — Physics of Transient Phenomena During Pulsed Laser Annealing and Sputtering. In: Ref. 302, p. 95.

305. Siegman A.E. — IEEE J. Q. E., 1986, QE-22, p. 1384.

306. Emel'yanov V.I., Sumbatov A.A. — Poverhnost, 1988, N 7, p. 122.

307. Emel'yanov V.I., Uvarova I.F. JETP, 1988, v. 94, p. 255.

Appendix

The fields in the medium (3.141, 3.142) can be represented in the form (see [275]):

$$E_x(x, y, z, t) = \exp\left[-\gamma(z - \xi(x, y))\right] \cdot$$

$$\sum_{a=0,\pm1} \varphi_a \exp\left[i k_a r - i a \varphi - i \omega t\right] + c.c.$$

$$E_y(x, y, z, t) = \exp\left[-\gamma(z - \xi(x, y))\right] \cdot$$

$$\sum_{a=0,\pm1} \psi_a \exp\left[i k_a r - i a \varphi - i \omega t\right] + c.c. \cdot$$

where $r_{pp} \approx r_{py}$; $\quad \varphi_0 \approx \dfrac{2kz}{k_z + i\gamma} E_{ix}$, $\quad \varphi_0 \approx -\dfrac{k_z}{k_0} \dfrac{2i\gamma}{\epsilon k_z + i\gamma} E_{i0}$,

$$T_p \approx (k_0 a_q)^2 \frac{k_0}{k_z} \frac{\epsilon k_0}{\epsilon \Gamma_p + \gamma}$$

$$\varphi_a \approx \frac{2i\gamma k_{ax}}{\epsilon \Gamma_a + \gamma} \frac{r_{ax} E_{ix} + r_{ay} E_{i0}}{1 - i \sum_{p=\pm1} T_p(r_{px}^2 + r_{py}^2)} a_q$$

$$\psi_a \approx \frac{2i\gamma k_{ay}}{\epsilon \Gamma_a + \gamma} \frac{r_{ax} E_{ix} + r_{ay} E_{i0}}{1 - i \sum_{p=\pm1} T_p(r_{px}^2 + r_{py}^2)} a_q$$

$$E_{i0} = k_0 E_{iz}/k_t, \quad r_{ax} = k_z k_{ax}/k_0^2, \quad r_{ay} = (k_t - k_{ay})/k_0,$$

$$\Gamma_a = (k_a^2 - k_0^2)^{1/2}, \quad \gamma = \sqrt{-\epsilon}\ k_0.$$

These expressions are used in (3.181).

TEUBNER-TEXTE zur Physik

Progress in Polymer Spectroscopy

Proceedings of the 7th European Symposium on Polymer Spectroscopy,
15 - 18 October 1985

Ed. by W. E. STEGER, Dresden
388 pp. (Vol. 9). Boards 48,50 M

ISBN 3-322-00344-2
Order No. 666 339 5 / Steger, Spectroscopy engl.

This Teubner-Text contains most of the lectures presented at the 7th
European Symposium on Polymer Spectroscopy ("ESOPS 7") held at the
University of Technology in Dresden, October 15th to 18th 1985. The
contents comprise the applications of advanced methods from ultra-
sonic to positron annihilation spectroscopy (and of related scattering
processes) with broadest coverage of magnetic resonance and optical
methods to the problems of analysis and structure research, stability
and processing of high polymers, mostly such of practical importance.

LEIPZIG   BSB B. G. TEUBNER VERLAGSGESELLSCHAFT

TEUBNER-TEXTE zur Physik

SOS 84

Proceedings of the Third Symposium Optical Spectroscopy
held in Reinhardsbrunn, 26 - 28 September 1984

Ed. by D. FASSLER, K.-H. FELLER and B. WILHELMI, Jena
398 pp. (Vol. 4) Boards 49,- M

ISBN 3-322-00656-5
Order No. 666 295 8 / Fassler, SOS 84 engl.

This Teubner-Text contains the text of lectures given at the III.
Symposium Optical Spectorscopy held at Reinhardsbrunn in September
1984. Contributions are given to the main topics of the conference:

A) New developments and applications of laser spectroscopy,including
    - time resolved UV/VIS-spectroscopy
    - time resolved fluorescence spectroscopy
    - laser Raman spectroscopy

B) Dynamics and photokinetics of molecular systems

C) Spectroscopy and photoprocesses in organized and biological
    systems

LEIPZIG   BSB B. G. TEUBNER VERLAGSGESELLSCHAFT

TEUBNER-TEXTE zur Physik

UPS - 85

Proceedings of the IV International Symposium Ultrafast Phenomena in
Spectroscopy
held in Reinhardsbrunn (GDR), October 23 - October 26, 1985

Ed. by E. KLOSE, Berlin, and B. WILHELMI, Jena
312 pp. (Vol. 10). Boards 39,- M

ISBN 3-322-00413-9
Order No. 666 352 0 / Klose, UPS 85

This Teubner-Text contains the text of lectures given at the IV International Symposium on Ultrafast Phenomena in Spectroscopy held in Reinhardsbrunn (GDR) in October 1985.
Contributions are given to the main topics of the conference:

I) New devices and methods:
    Pico- and femtosecond pulses, high-power pulses, UV- and IR-pulses,
    nonlinear optical devices.

II) Applications:
    Application to solid state physics, opto-electronics, molecular
    physics, chemical reactions, biological processes.

BSB B. G. TEUBNER VERLAGSGESELLSCHAFT